STRUCTURAL PACKAGE DESIGNS

STRUCTURAL PACKAGE DESIGNS

VERPACKUNGSFORMGEBUNG • MODÈLES STRUCTURAUX DE CONDITIONNEMENT
DISEÑOS DE ESTRUCTURAS PARA EMBALAJES • DESIGN STRUTTURALE DELLA CONFEZIONE
包裝結構設計 • パッケージの立体デザイン

Haresh Pathak

THE PEPIN PRESS

AMSTERDAM AND SINGAPORE

Copyright for this edition © 1998, 1999 The Pepin Press B/v

ISBN 90 5496 051 5

A catalogue record for this book is available from the publishers
and from the Dutch Royal Library, The Hague

All designs by Haresh Pathak, Bombay
Originally published in 1996 by Super Book House, Bombay
This book is edited, designed and produced by
The Pepin Press in Amsterdam and Kuala Lumpur
Cover design and typography: Pepin van Roojen
Copy-editing introduction: Andrew May
Translations: Sebastian Viebahn with assistance from Jutta Riedel (German);
LocTeam (Spanish); Anne Loescher (French); Luciano Borrelli (Italian);
Mitaka (Chinese and Japanese)

The Pepin Press
P.O. Box 10349
1001 EH Amsterdam
The Netherlands
Tel (+) 31 20 4202021
Fax (+) 31 20 4201152
mail@pepinpress.com
www.pepinpress.com

Printed in Singapore

7 8 9 / 02 01

Contents

Structural Package Designs

Packaging is an important factor in any retail environment and a key element in most marketing strategies. Consumers react immediately to package shapes, and are influenced by them when making buying decisions. Different product categories are often easy to recognize by their characteristic form, for example chocolate boxes or milk cartons.
On the other hand, a manufacturer of an exclusive product, such as jewellery or perfume, may deliberately choose an unusual, eye-catching form.
Paper-board is the most used material in packaging. Some reasons for its popularity are that it is relatively cheap and easy to produce and process and, after use, it is easy to recycle. Most printing techniques also give excellent results on paper-board.
This book serves as a reference for structural design. All designs have been selected on account of their functional relevance and acceptability, and can be easily modified to suit specific requirements.

Verpackungsformgebung

Verpackungen sind ein wichtiger Faktor in sämtlichen Bereichen des Einzelhandels und ein Schlüsselelement der meisten Marktstrategien. Konsumenten reagieren unmittelbar auf die Form einer Verpackung und werden dadurch in ihren Kaufentscheidungen beeinflusst. Durch typische Erscheinungsbilder, zum Beispiel Schokoladenverpackungen oder Milchkartons, lassen sich oftmals auf einfache Art bestimmte Produktkategorien definieren. Hersteller exklusiver Produkte wie Schmuck oder Parfüms können sich aber auch bewußt für ungewöhnliche Formen entscheiden. Das bei der Verpackungsherstellung meistverwendete Material ist Pappe. Sie ist unter anderem deshalb so beliebt, weil sie relativ preiswert, einfach herzustellen und leicht zu bearbeiten ist, und sich nach Gebrauch problemlos wiederverwerten läßt. Zudem liefern die meisten Druckverfahren auf Pappe ausgezeichnete Ergebnisse.

Das vorliegende Buch ist ein Referenzwerk für die Verpackungsformgebung. Ausschlaggebend bei der Auswahl der Designs waren Akzeptanz und funktionelle Bedeutung. Zur Anpassung an spezifische Anforderungen lassen sich die Designs problemlos variieren.

Diseños de estructuras para embalajes

El embalaje es un factor primordial en cualquier entorno comercial y un elemento clave en gran parte de las estrategias de mercado. Los consumidores reaccionan de manera inmediata ante la forma de los envases, lo que les influirá a la hora de decidir qué comprar. Existen modelos muy característicos que a menudo definen categorías de productos, como es el caso de las cajas de bombones o los cartones de leche. En cambio, el fabricante de un producto exclusivo, como joyas o perfumes, tiene la opción de decantarse por una forma poco habitual. El cartón es el material más utilizado en los embalajes. Algunas de las razones de esta popularidad son el hecho de que resulta relativamente económico, la escasa complicación que presenta su producción y manipulación, así como la posibilidad de reciclarlo fácilmente después de su uso. Por otro lado, la mayoría de las técnicas de impresión ofrecen excelentes resultados sobre dicho material.

Este libro sirve como referencia para la realización de diseños de estructuras. Todos ellos han sido seleccionados en razón de su aceptabilidad y relevancia funcional y, además, pueden modificarse fácilmente a fin de adaptarlos a necesidades concretas.

Design strutturale della confezione

La confezione è un'importante fattore in ogni ambiente di vendita e un elemento chiave della maggioranza delle strategie di marketing. Il consumatore reagisce immediatamente alla forma della confezione ed è da questa influenzato nel momento della scelta. Alcune categorie di prodotti sono facilmente riconoscibili dalla confezione grazie alla sua forma caratteristica come ad esempio le scatole di cioccolatini o i cartoni del latte. Per alcuni prodotti esclusivi invece, come profumo o gioielli, il produttore potrebbe deliberatamente optare per una forma più inusuale ed appariscente.

Il cartone è di gran lunga il materiale più usato per la produzione di confezioni. Alcune ragioni di questo successo sono i costi relativamente ridotti, la facilità di produzione e di lavorazione e, dopo l'uso, la reciclabilità. Inoltre, la maggior parte delle tecniche grafiche utilizzate si applicano ottimamente a questo materiale.

Il presente volume servirà come consultazione relativamente al design strutturale. Tutti i progetti sono stati selezionati in base alla loro rilevanza ed accettabilità strutturale e possono essere facilmente adattati ad esigenze specifiche.

Modèles structuraux de conditionnement

Le conditionnement constitue un facteur important dans tout environnement commercial et est un élément clé dans la plupart des stratégies de marketing. En effet, les consommateurs sont immédiatement sensibles aux formes des emballages et en seront influencés par elles dans leurs décisions d'achats. Souvent les aspects extérieurs du produit définissent le type de produit contenu, comme par exemple les boîtes de chocolat ou les cartons de lait. Par ailleurs, un fabricant de produits exclusifs, comme le parfum ou la bijouterie, pourra délibérément opter pour une forme inhabituelle de conditionnement. Pour diverses raisons, le carton est le matériel le plus couramment utilisé pour les emballages. Cette popularité est notamment due à l'aspect relativement peu onéreux du matériel, à sa facilité d'utilisation et de recyclage aprés usage. De plus, la plupart des techniques d'impression offrent d'excellents résultats sur le carton.

Ce livre sert de point de référence pour la création des structures en volume. Tous les modèles représentés ont été sélectionnés selon leur capacité fonctionnelle et leur acceptabilité et peuvent être facilement modifiés pour être adaptés à des besoins spécifiques.

包裝結構設計

包裝在任何零售行業中都起著重要的作用，也是市場戰略中的關鍵要素。消費者對商品的包裝十分敏感，包裝的好壞將影響到消費者是否下購買的決心。特有的包裝外表使我們常常容易地辨別出商品的種類，比如巧克力盒和牛奶盒。而另一方面，一個高檔商品如首飾或香水的製作人可能會故意選擇一種不同尋常的包裝形式。

紙板是用作包裝的主要材料。其原因是紙板相對比較便宜，容易生產和加工，使用之後容易回收利用。而且，大部份印刷技術在紙板材料上能夠獲得最佳效果。

本書為包裝結構設計提供參考。書中所有的設計都是根據其實用價值和能夠被接受採用而選擇的，而且容易修改以適應特殊的要求。

パッケージの立体デザイン

パッケージングは小売商品にとって非常に重要な要素であり、マーケティングでのキーポイントとなることが少なくありません。消費者にすぐアピールするのがパッケージの形であり、買うかどうかの決め手にもなるでしょう。普通、外見の特色で商品の種類がすぐにわかります。チョコレートの箱や牛乳パックなどがその良い例です。逆に、宝石や香水などの高価な商品のメーカーは、わざと変わった形を選ぶこともあります。パッケージ材として最も広く使われているのは板紙です。比較的安価であること、生産が簡単で作業もしやすいこと、そして使用後はリサイクルできることなどが人気の秘密です。さらに、この素材ならほとんど全ての印刷技術で優れた結果が得られます。

本書を立体デザインの参考としてご利用ください。デザインは全て機能面と受容性を基準にセレクトしてあり、特殊な要求に合わせて応用しやすくなっています。

Basic Structures

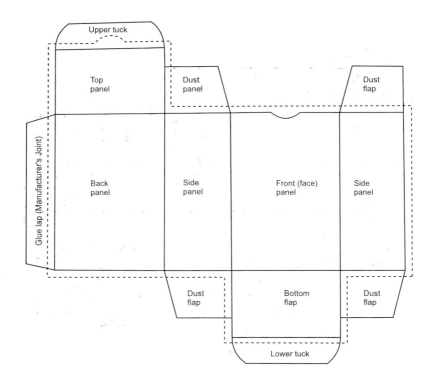

--- Indictes typical limit for printing (bleed)

Terminology for Folding Cartons

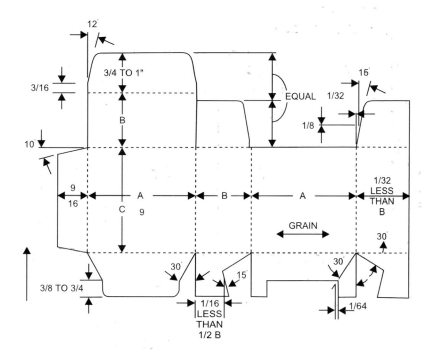

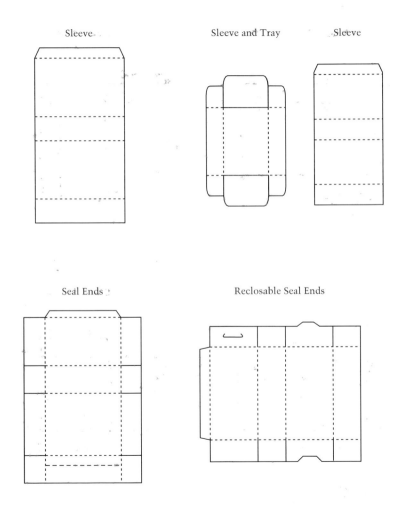

Sleeve

Sleeve and Tray

Sleeve

Seal Ends

Reclosable Seal Ends

Basic Structures

Straight Tuck

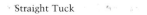

Reverse Tuck

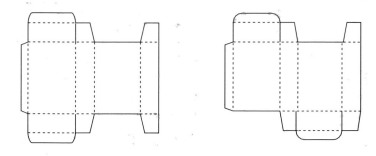

Tapered Top

Window Carton

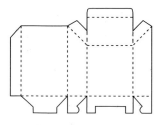

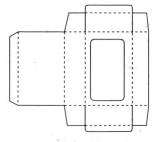

Tray

Two Piece Tray and Lid

Hollow Wall Tray

Display Package

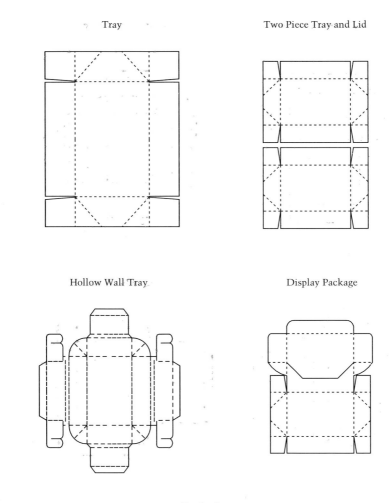

Basic Structures

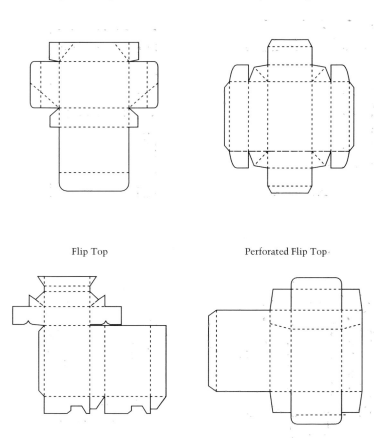

Tray with Hinged Lid

Tray with Hinged Lid

Flip Top

Perforated Flip Top

Dispensing

Spout

Universal

Full Overlap

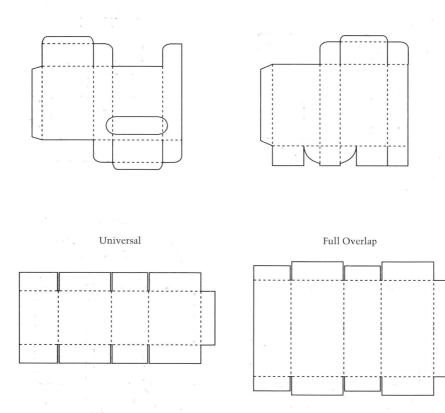

Basic Structures

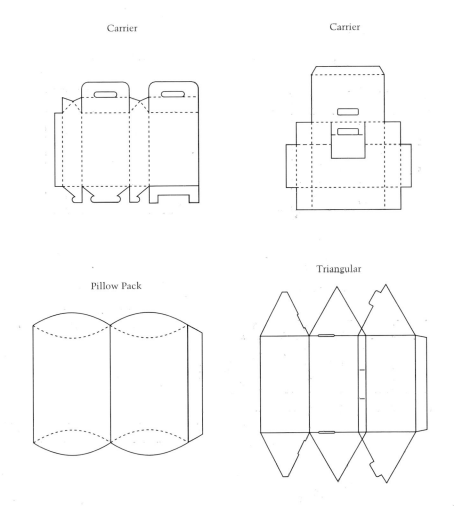

Carrier

Carrier

Pillow Pack

Triangular

Basic Structures

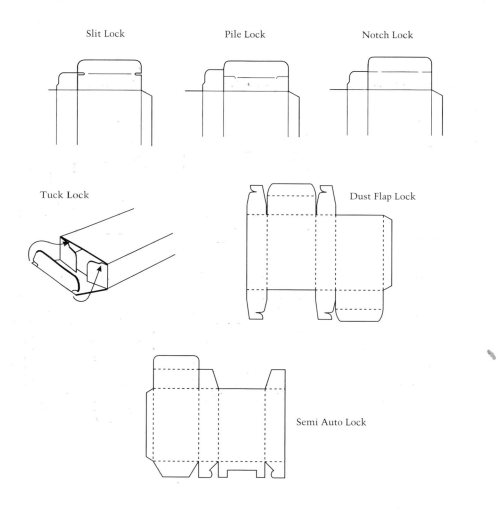

Slit Lock

Pile Lock

Notch Lock

Tuck Lock

Dust Flap Lock

Semi Auto Lock

Locking Methods

Gusseted Dust Flaps

Top Panel Lock

Top Panel Lock

Semi Auto Lock

Auto Lock

Top Panel Lock

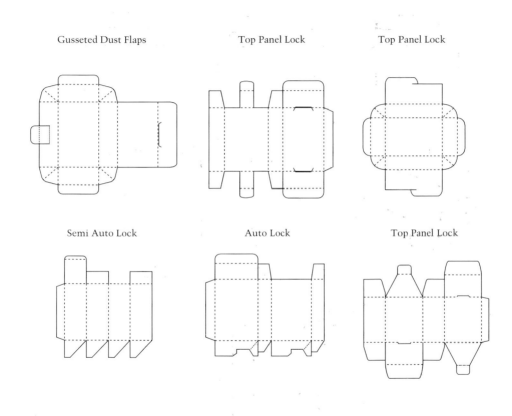

Top Panel Locks   Side Panel and End Panel Locks

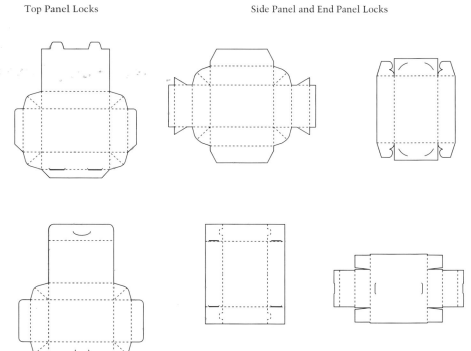

Regular Folding Boxes

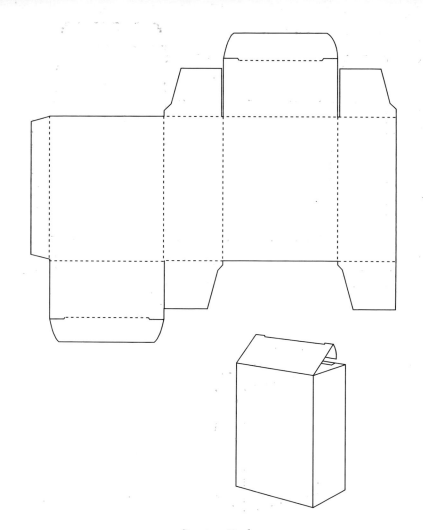

Reverse Tuck

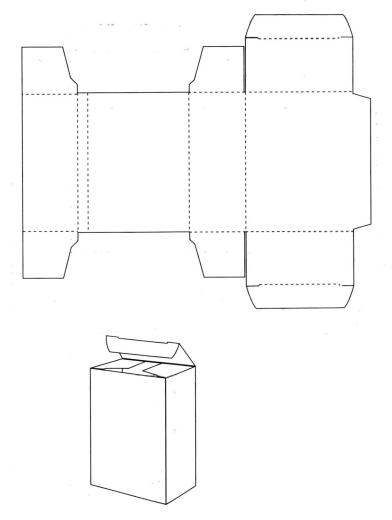

Straight Tuck

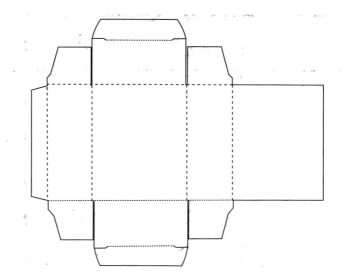

Straight Tuck with Slit Lock

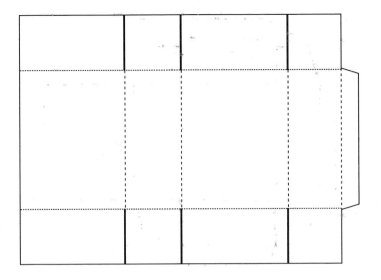

Full Overlap Seal Ends

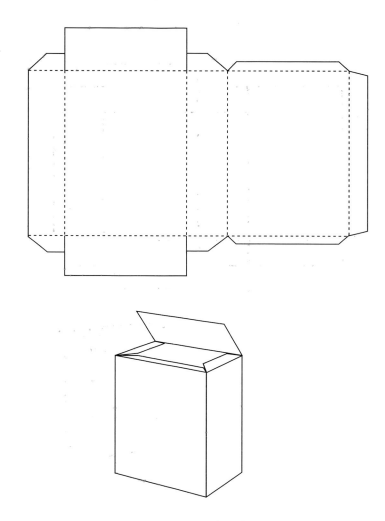

Economy Seal Ends

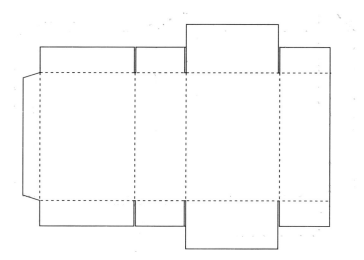

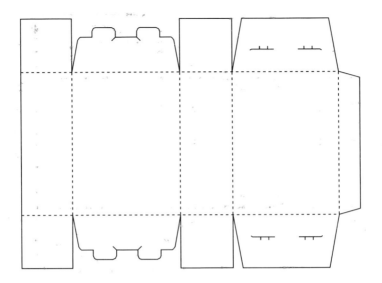

Partial Overlap Tuck Lock Flaps

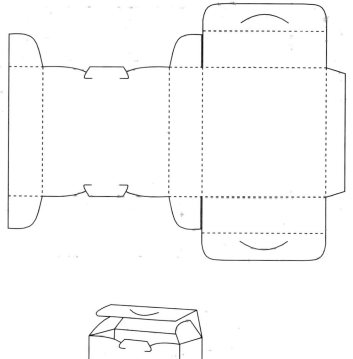

Tuck with Anchor Lock

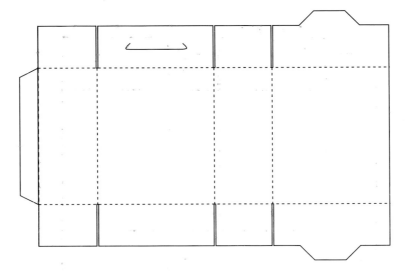

Partial Overlap Seal Ends with Dispenser Tab Lock

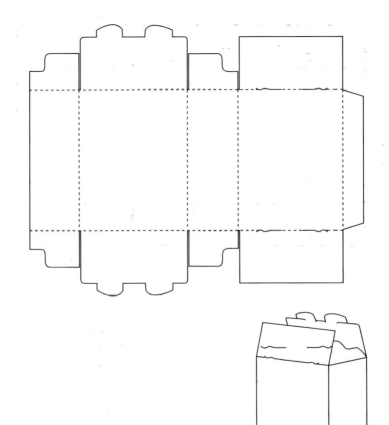

Full Overlap with Edge Lock

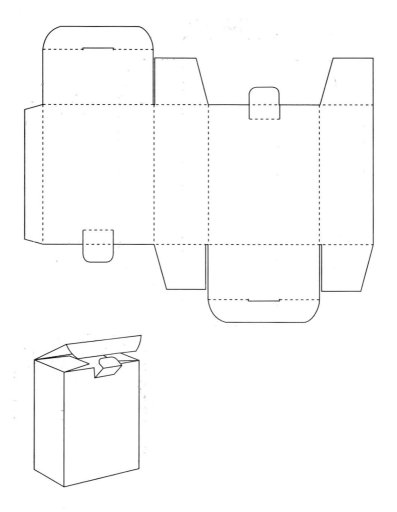

Tuck and Tongue

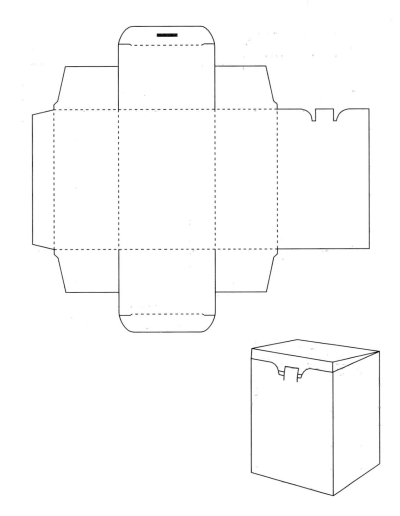

Tuck and Tongue

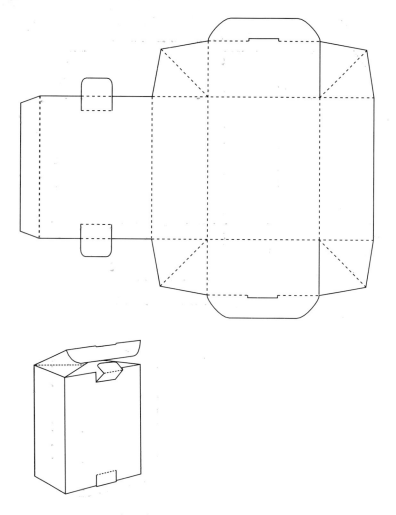

Tuck and Tongue with Gusset Dust Flap

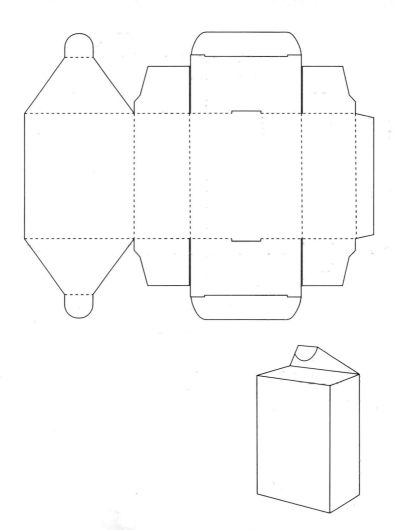

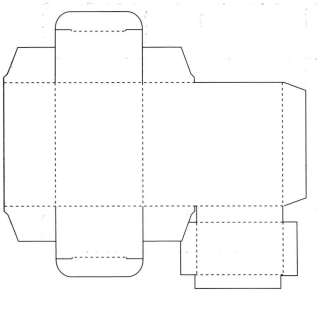

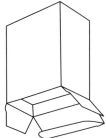

Straight Tuck with Inside Bottom Platform

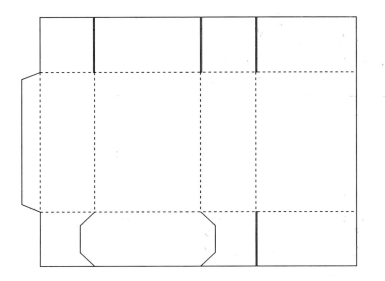

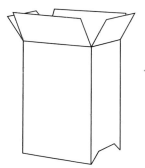

Full Overlap Seal End with Vanburen Ears at Bottom

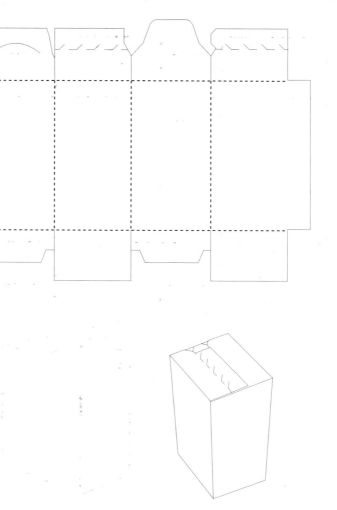

Seal End with Tab Lock and Zipper

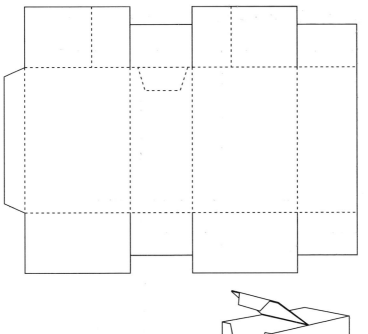

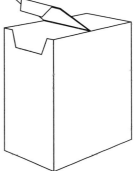

Overlap Seal Ends with Perforated Spout

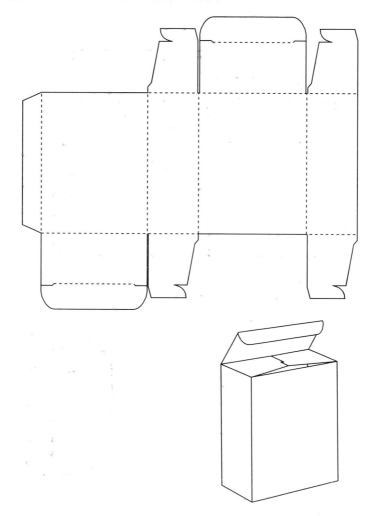

Reverse Tuck with Locking Dust Flaps

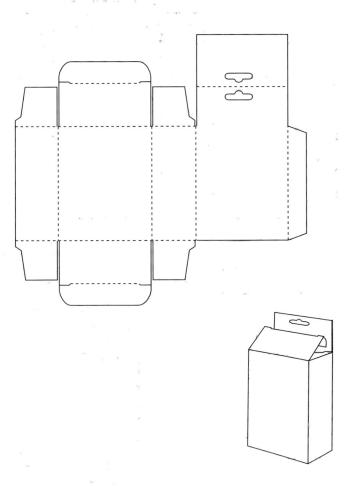

Straight Tuck with Hanging Panel   47

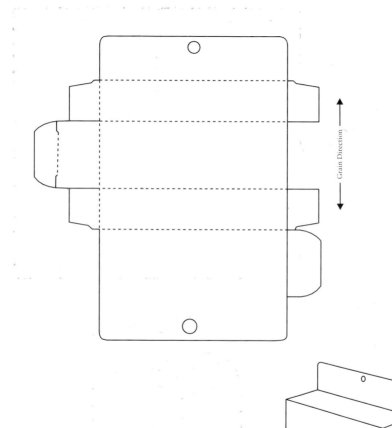

Grain Direction

Reverse Tuck with Header Card or Fifth Panel

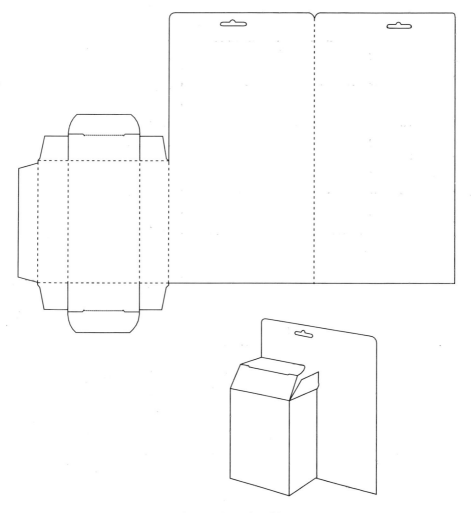

Straight Tuck with Fifth Panel

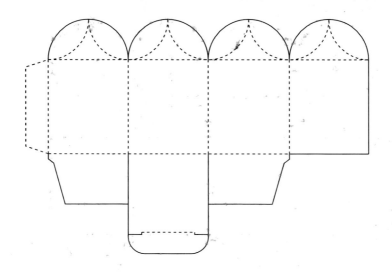

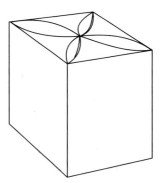

Top Closure Gift Lock

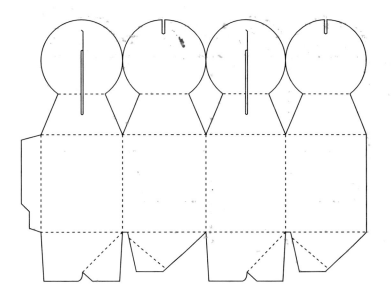

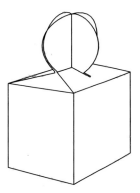

Fancy Top with Auto Lock Bottom

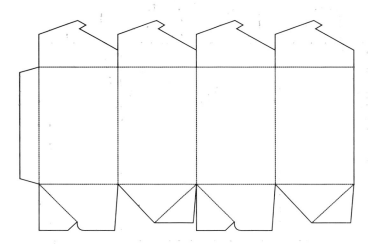

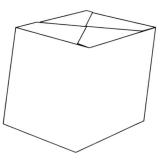

Altering Closure Top with Auto Lock Bottom

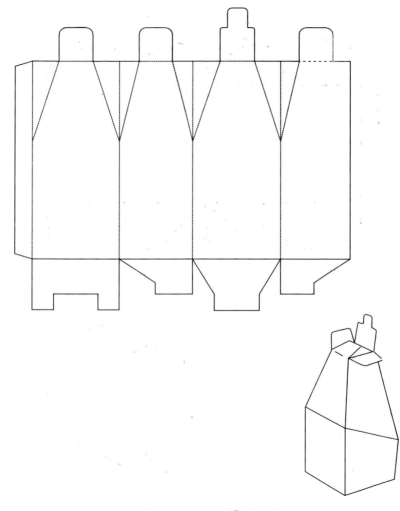

Square Dome Top

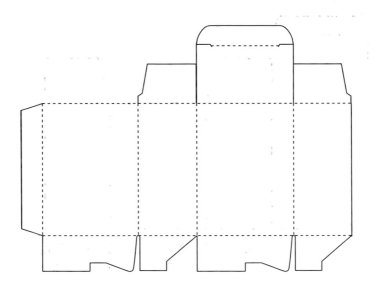

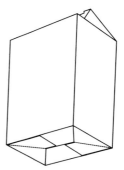

Tuck with Auto Lock Bottom

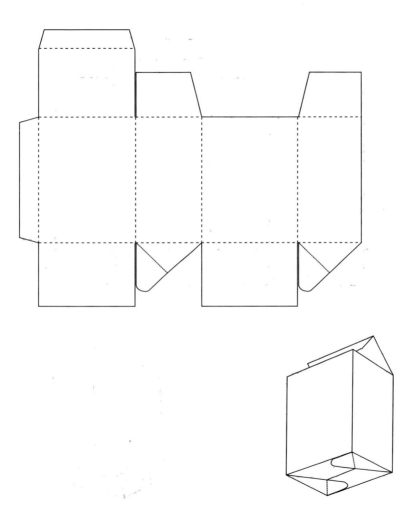

Tuck with Auto Lock Bottom

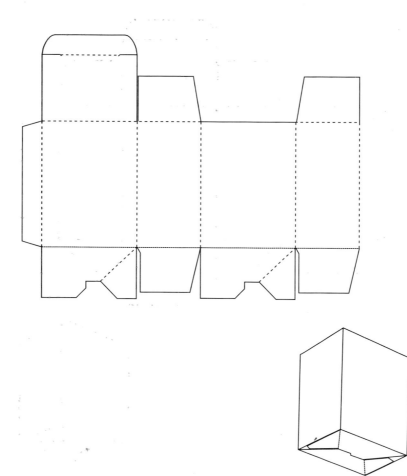

Tuck with Auto Lock Bottom

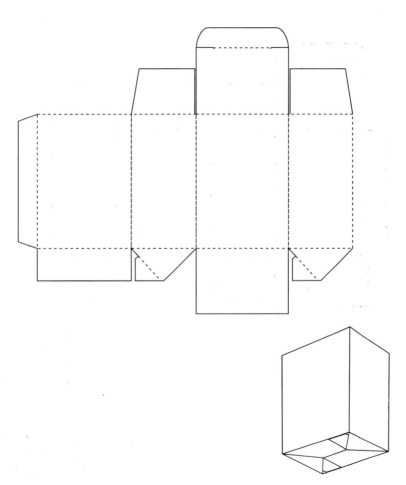

Tuck with Auto Lock Bottom

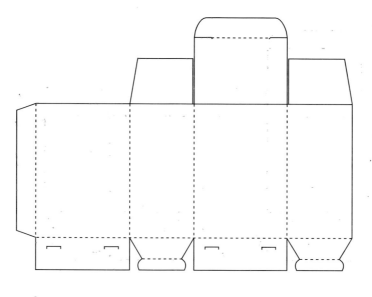

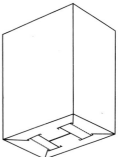

Ear Hook Double Lock Bottom

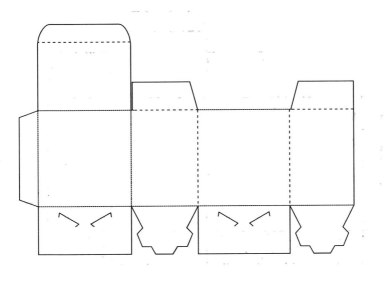

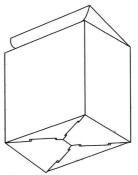

Ear Hook Double Lock Bottom

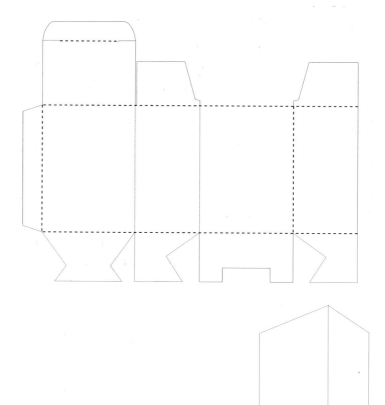

Reinforced Snap Lock Bottom

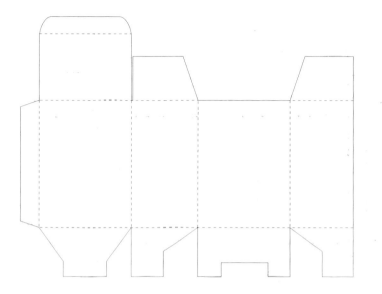

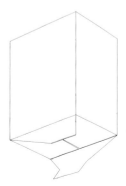

Snap Lock Bottom 61

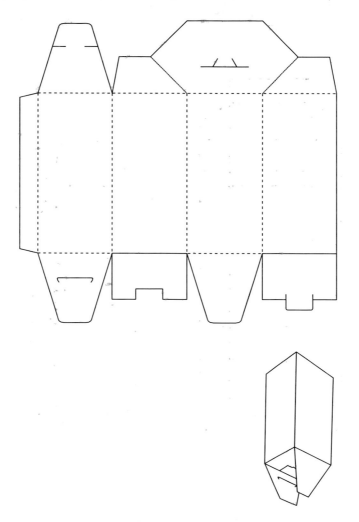

Double Snap Lock Top with Bellow Flap, Snap Lock Bottom

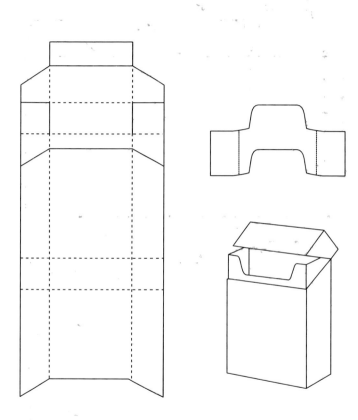

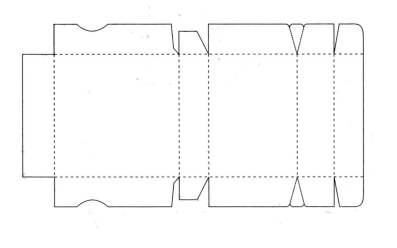

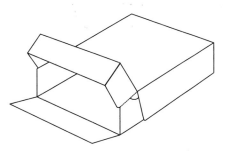

Butter Carton with Tear Away Tabs

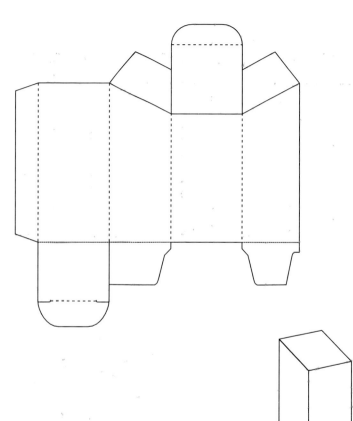

Reverse Tuck with Tapered Top

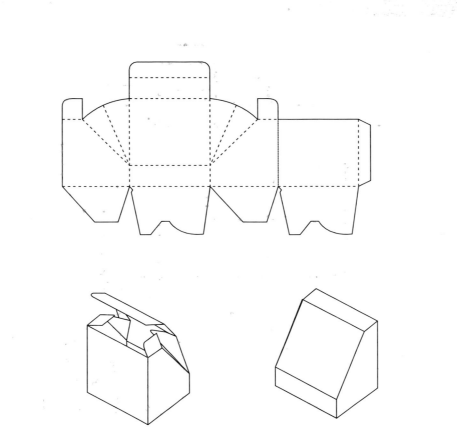

Faceted Top

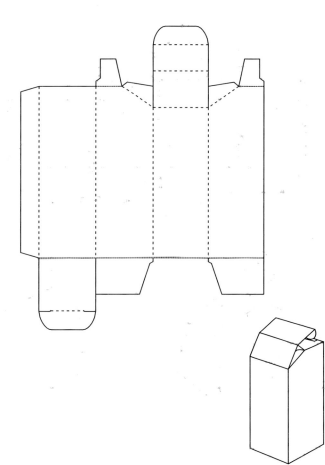

Faceted Front with Gusset Flap

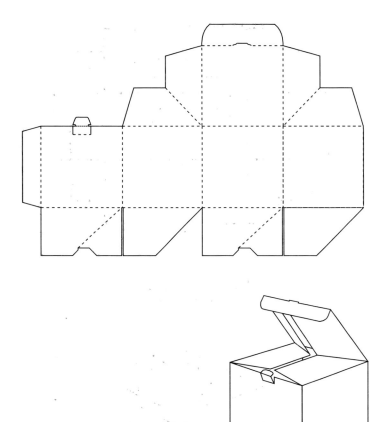

Bellow Dust Flap, Tuck Lock, Auto Lock Bottom

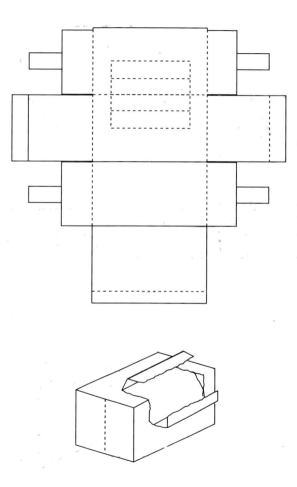

Straight Tuck Dust Flap with Roller Support and Perforated Top

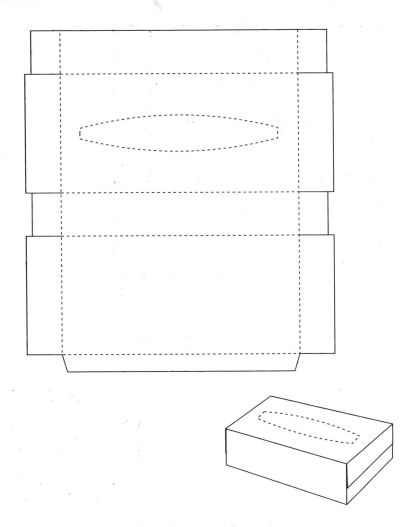

Seal End with Perforated Top (Tissue Dispenser)

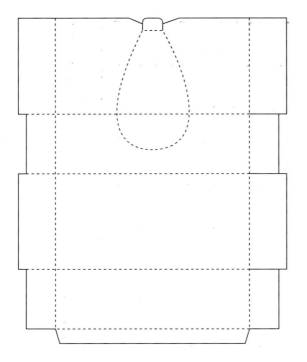

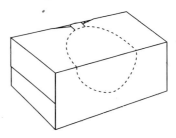

Seal End with Perforated Top and Front Panel　　71

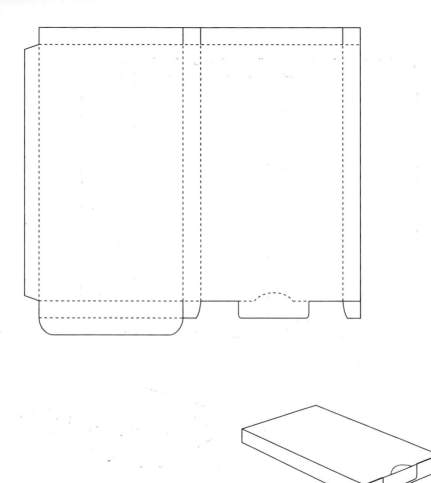

Tuck and Seal Flaps

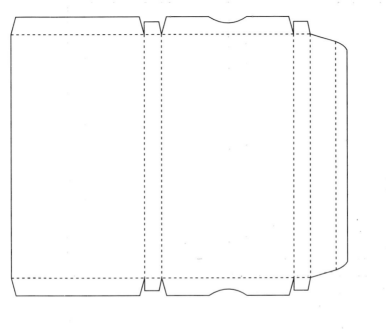

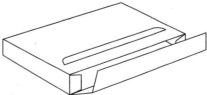

Side Seal End withg Tear Away Tuck

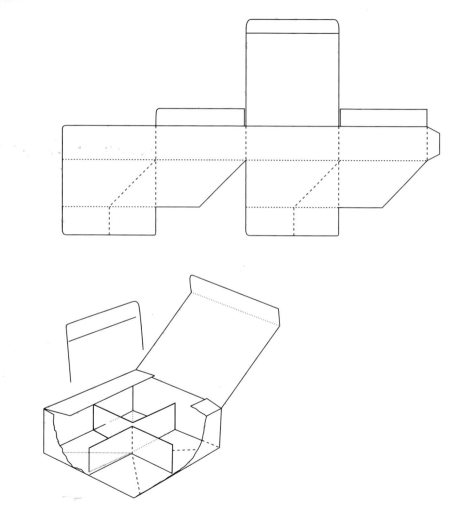

Tuck Top and Lock Bottom with Self Partition

Aseptic Packaging

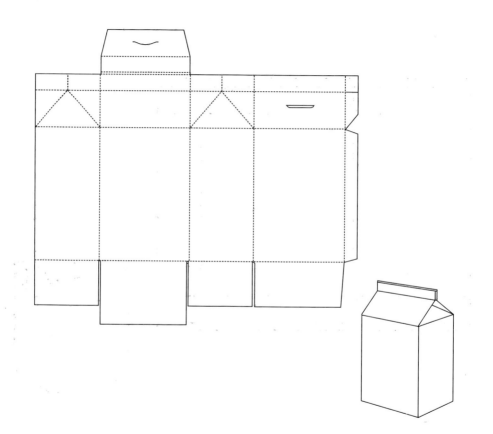

Milk Carton Style Container with Slit/Loop Closure

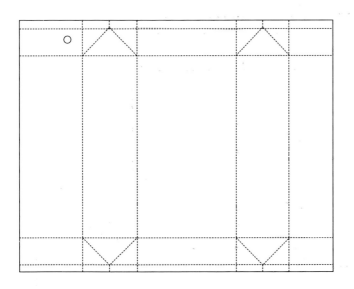

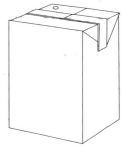

Aseptic Package

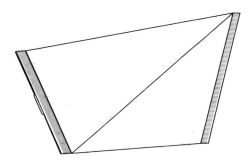

Tetrahedron for Aseptic Packaging of Liquid

Tray and Lid Boxes

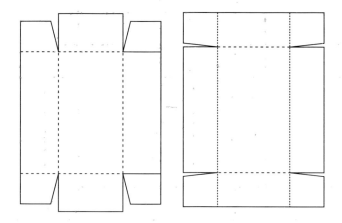

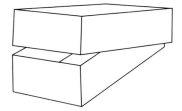

Two Piece Full Telescopic Tray

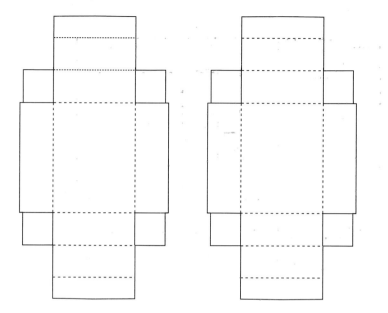

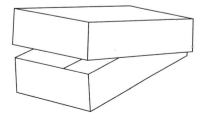

Two Piece Full Telescopic Friction Lock Tray

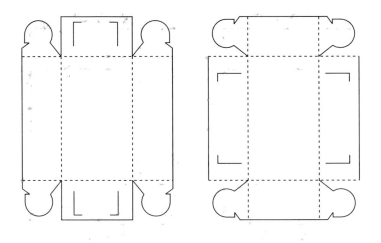

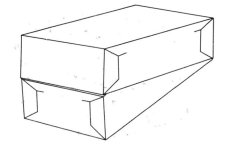

Full Telescopic Tray with Ear Hook Lock Corners

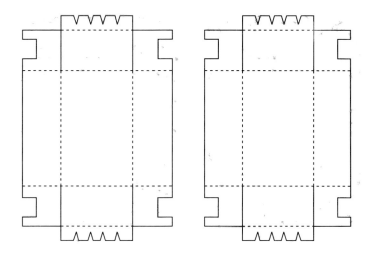

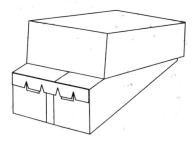

Two Piece Full Telescopic Tray and Lead With Seal Ends

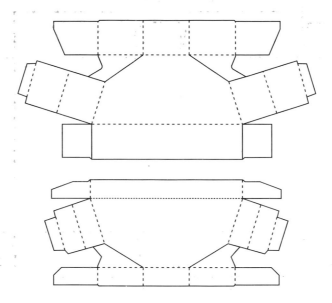

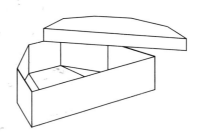

Self Locking Six Sides Tray and Partial Telescopic Cover

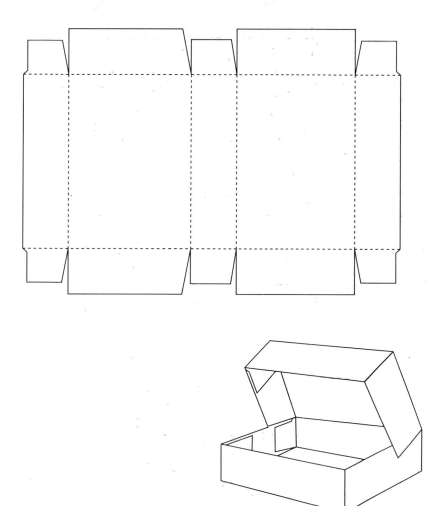

One Piece Tray/Lid with Glued Corners

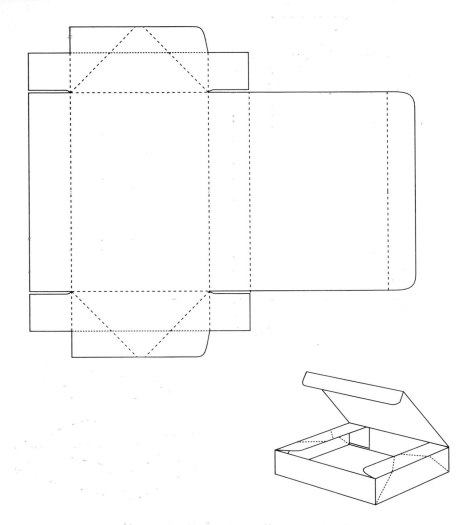

One Piece Collapsible Tray/Lid with Dust Flap

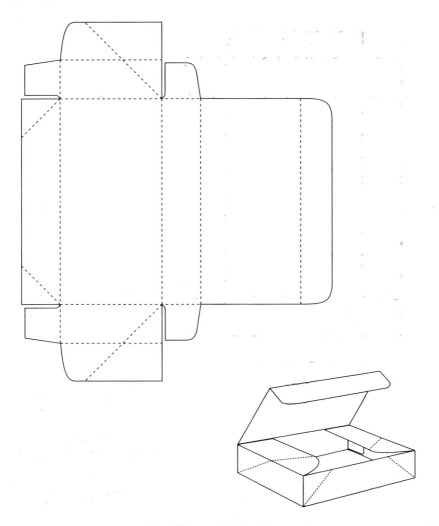

One Piece Collapsible Tray/Lid with Dust Flap

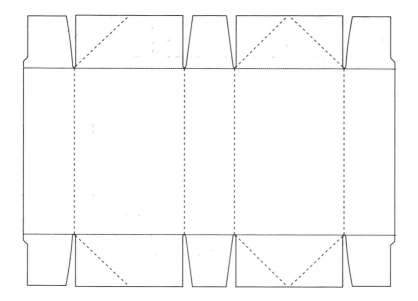

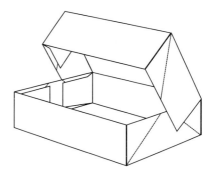

One Piece Collapsible Tray/Lid

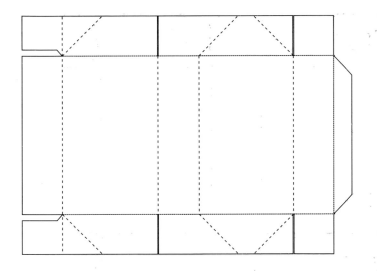

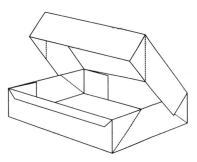

One Piece Collapsible Tray/Lid

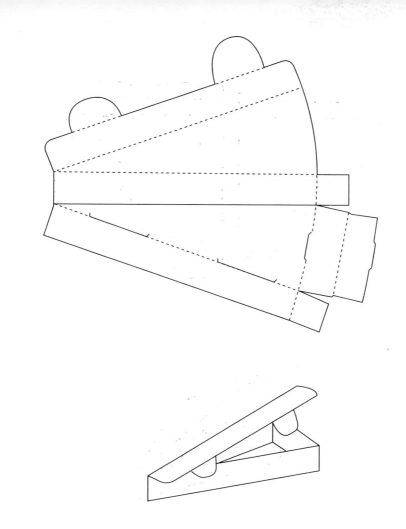

Self Corner Triangle with Tab Lock

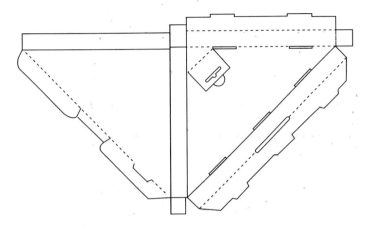

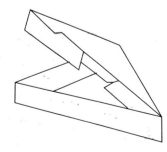

Triangular Self Locking Pegboard Hanger

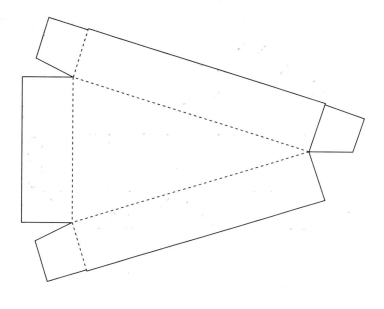

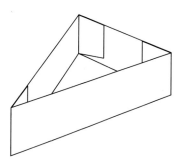

Triangular Tray with Glued Side Walls

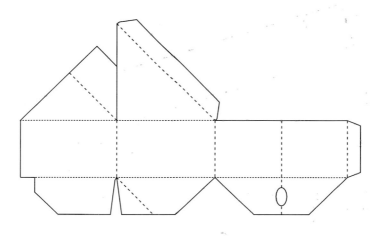

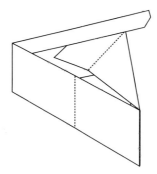

Triangular Folding Carton with Tuck Lid

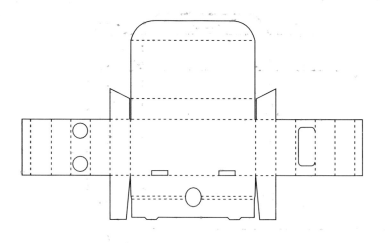

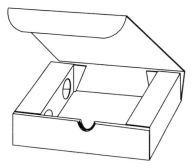

Self-Locking Tray/Lid with Built-in Supports

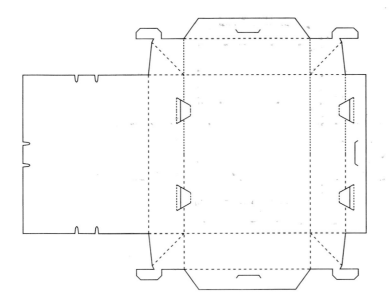

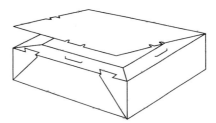

Self Locking Tray/Lid with Bellow Corners

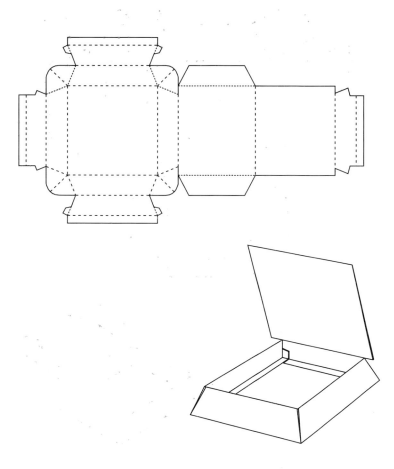

Self Locking Variation with Tapered Ends and Hollow Side Walls

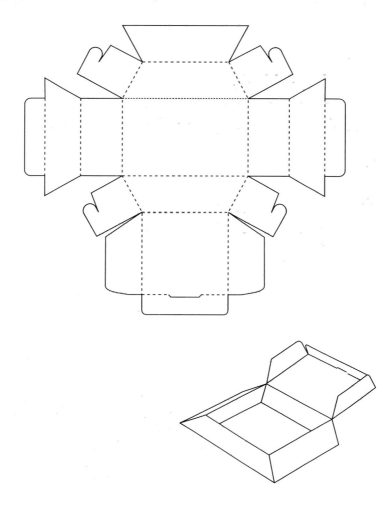

Tapered Ends and Hollow Side Walls with Hook Locks

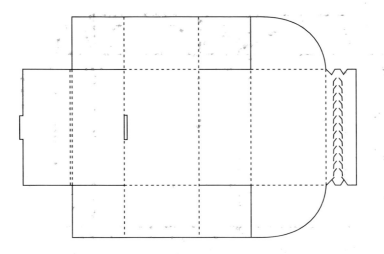

Self Locking Tray/Lid with Zip Tear Strip

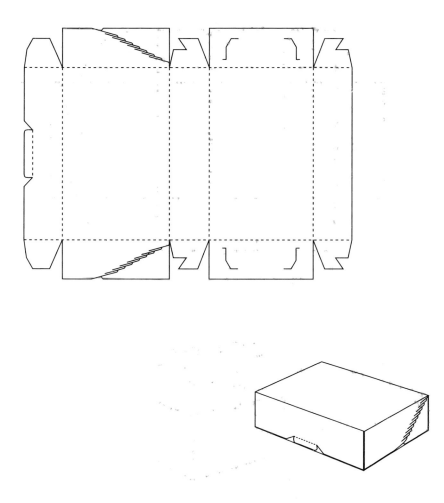

Self Locking Tray/Lid with Tear-away Side Seals

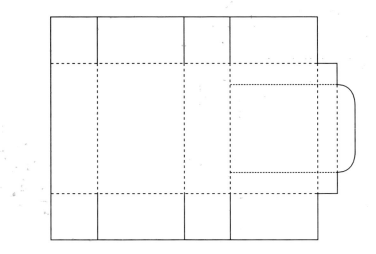

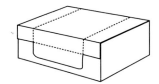

Seal End Style Tray with Perforated Top

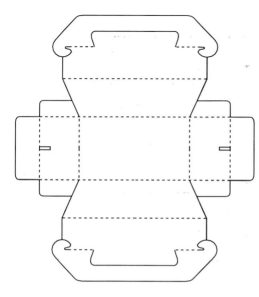

Trays

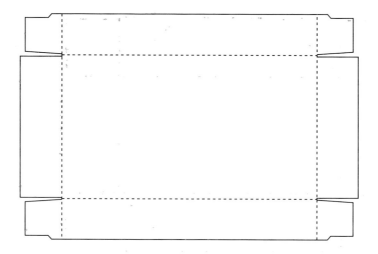

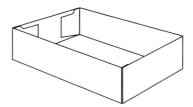

Side Seal Tray

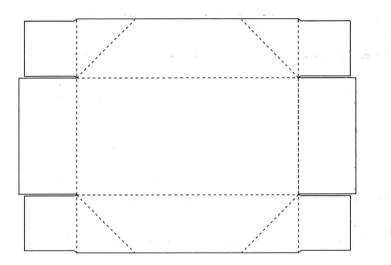

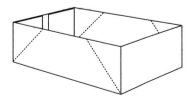

Side Seal Tray with Folding Side Wall

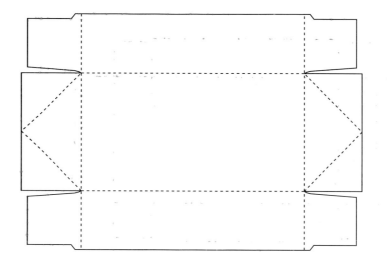

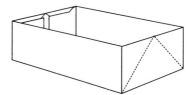

Side Seal Tray with Folding End Wall

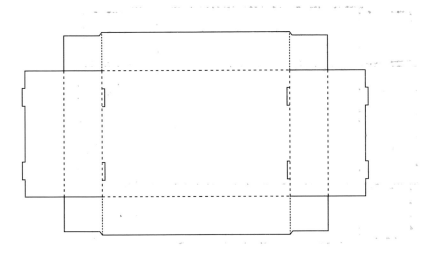

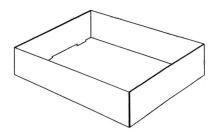

Double End Wall Self Locking Tray

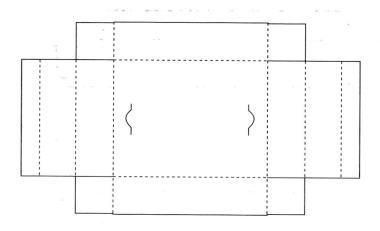

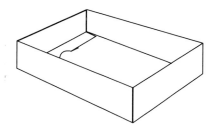

Double End Wall Tab Locking Tray

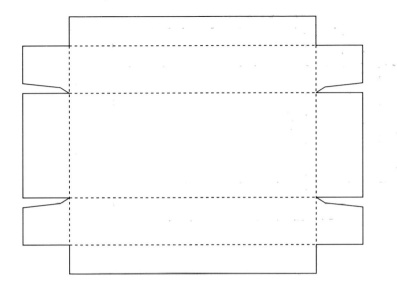

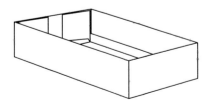

Glued Double Side Wall Tray

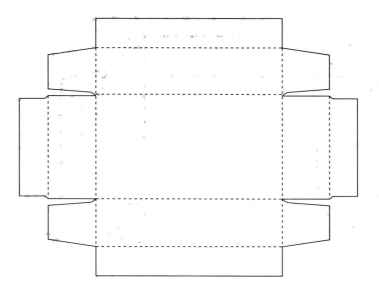

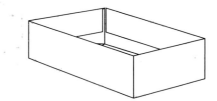

Partial Double Wall Tray with Glued Flaps

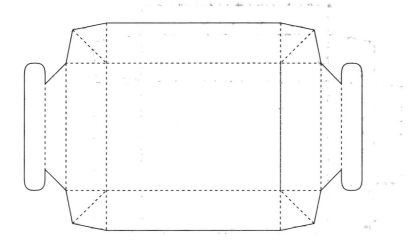

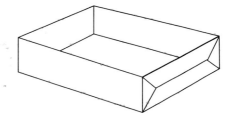

Outside T-Lock Tray

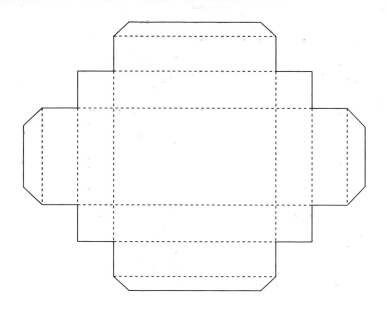

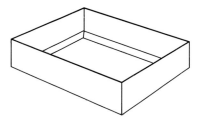

Foot Lock Double Wall Tray

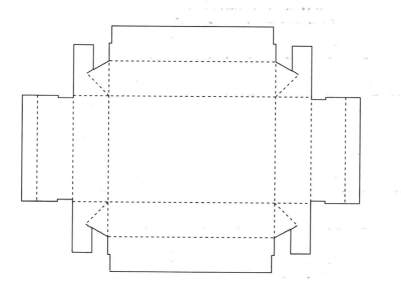

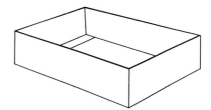

Self Locking Double End Walls and Glued Double Side Walls

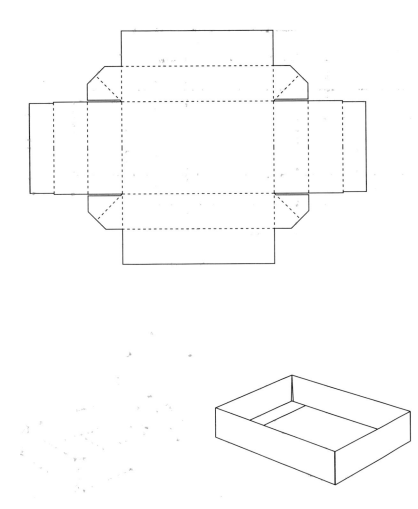

Self Locking Double Walls

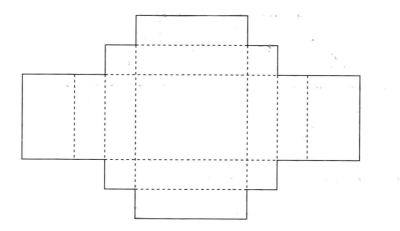

Glued Double Side Walls and Double Bottom

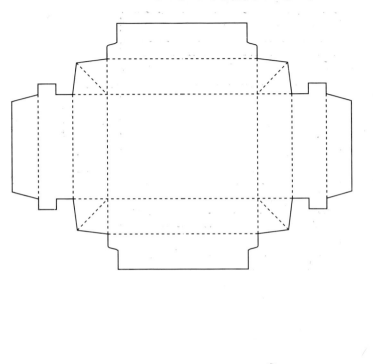

Self Locking Notch Lock Double Walls

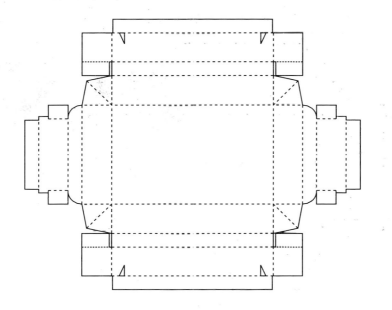

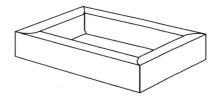

Self Locking Hollow Walls

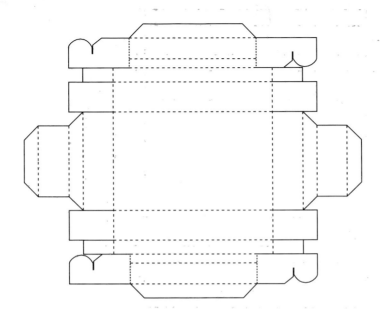

Self Locking Hollow Walls

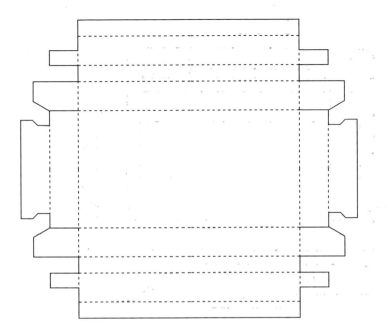

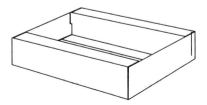

Self Locking Hollow Double Side Walls and Tapered End Walls 119

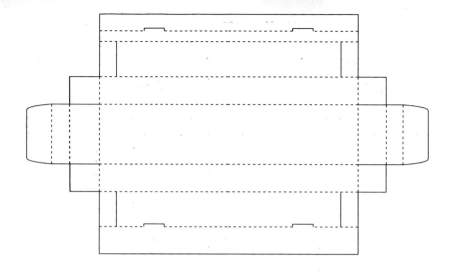

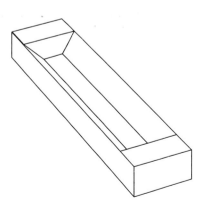

Self Locking Hollow Double End Walls and Tapered Side Walls

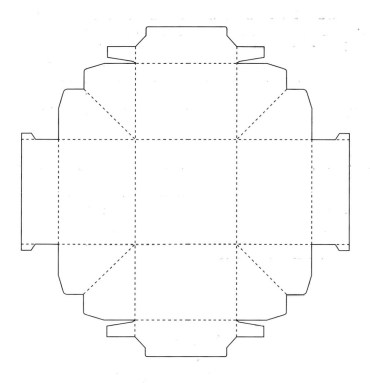

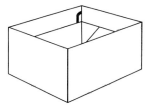

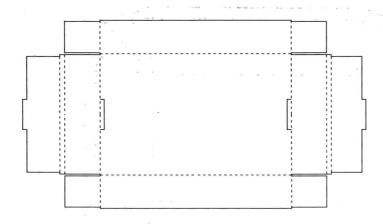

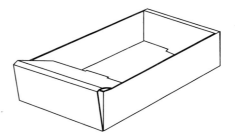

Self Locking Tray with Thick Double End Wall

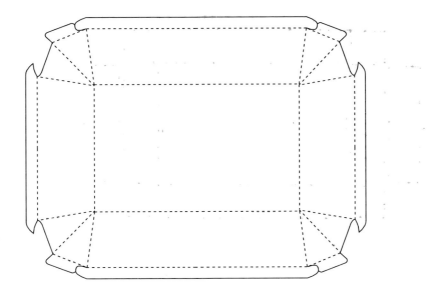

Glued or Heat-Sealed Tapered Tray with Fringe

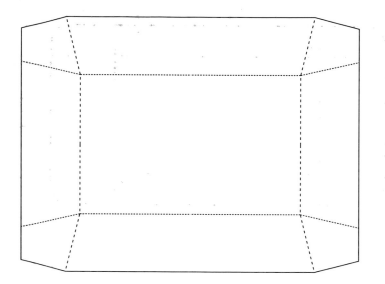

Tapered Tray with Glued Corners

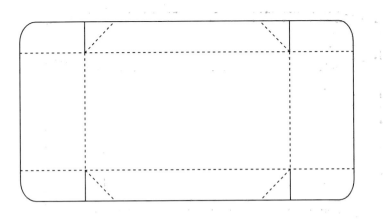

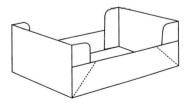

Tray with Raised End Panels and Sealed Corners 125

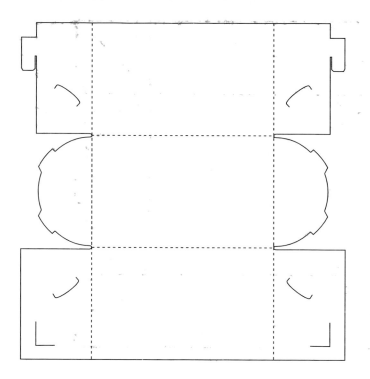

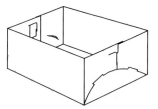

Double Lock Deep Tray

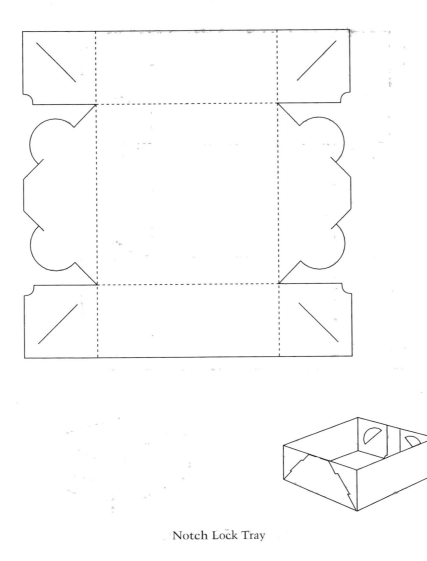

Notch Lock Tray

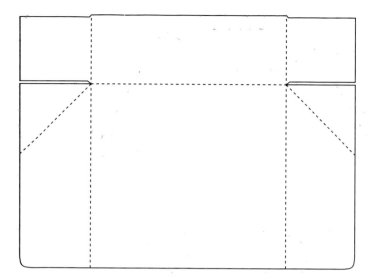

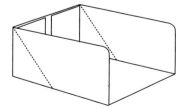

Open Ended Collapsible Tray with Two Sealed Corners

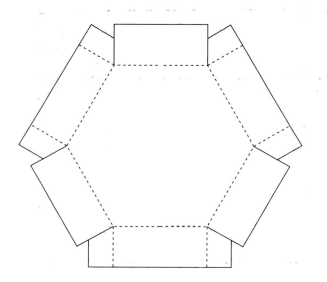

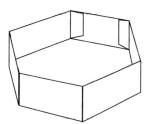

Hexagonal Tray

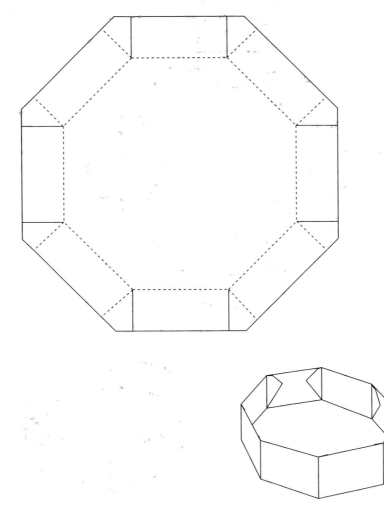

Octagonal Tray

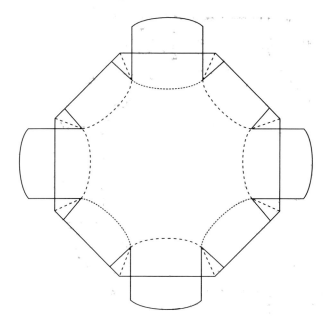

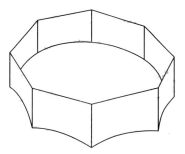

Octagonal Tray with Curved Sides

Sleeves

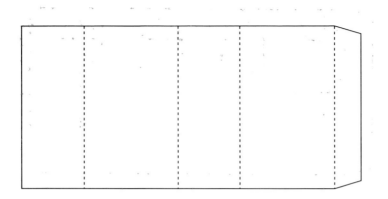

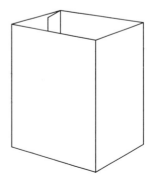

Simple Tube Sleeve

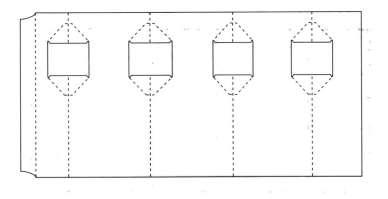

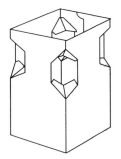

Sleeve with Built-in Cushion for Fragile Products 135

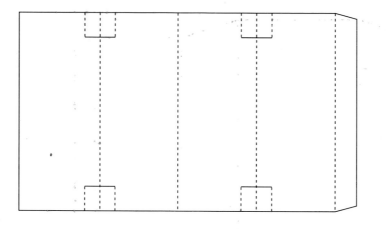

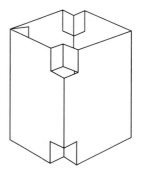

Tube Sleeve with Recessed Product Retainers

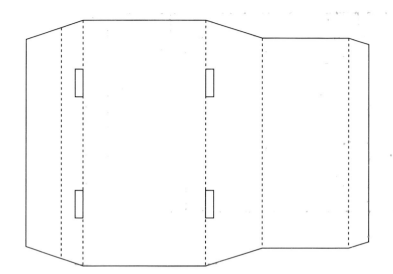

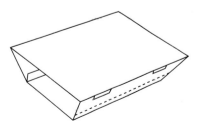

Tapered Sleeve with Glued Bottom

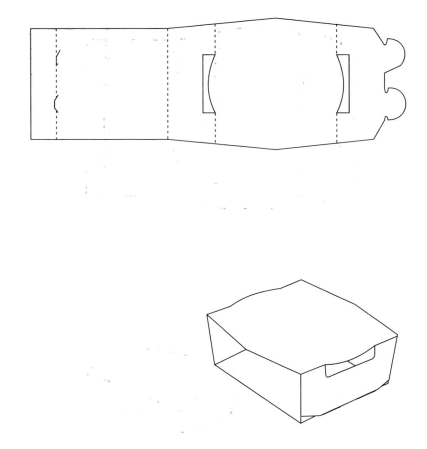

Edge-Locked Sleeve

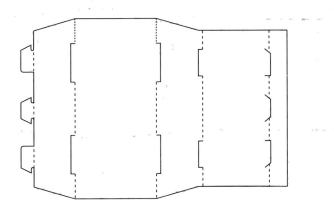

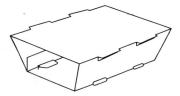

Edge-Locked Sleeve

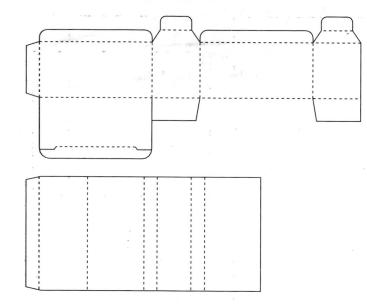

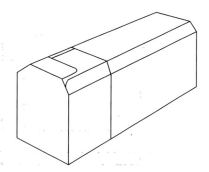

Tapered Top Sleeve with Fitting Tray

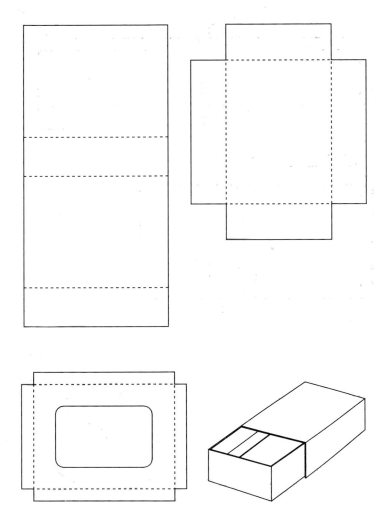

Tube Sleeve and Tray   141

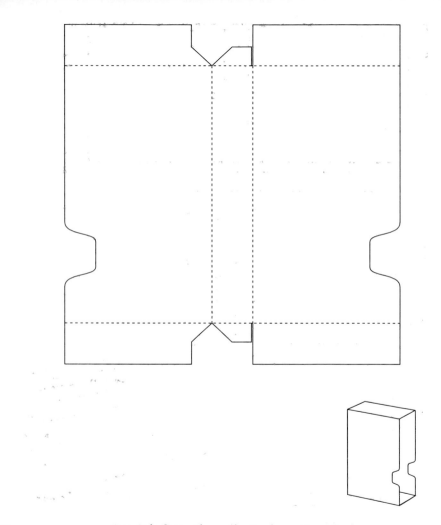

One Side Open Sleeve (for Books or Cassettes)

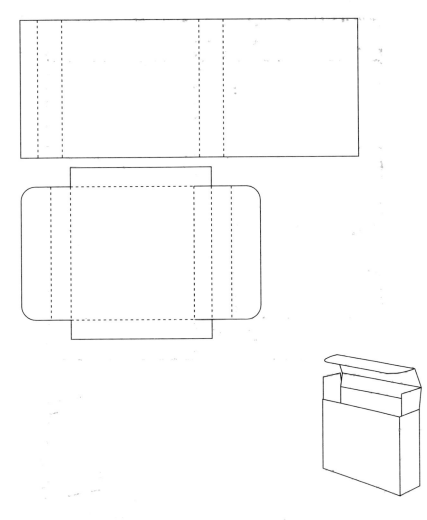

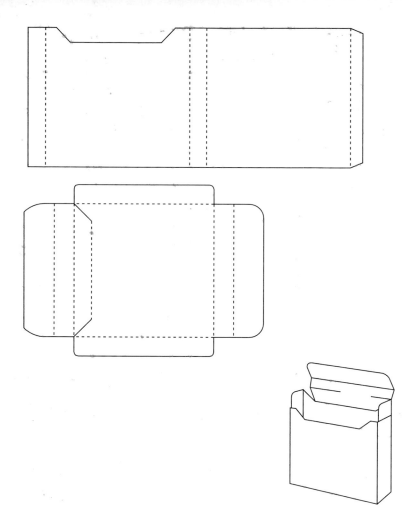

Two Piece Dispenser Sleeve/Tray

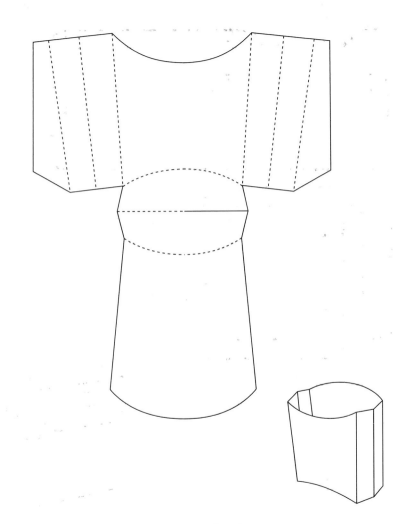

Container with Closed Bottom

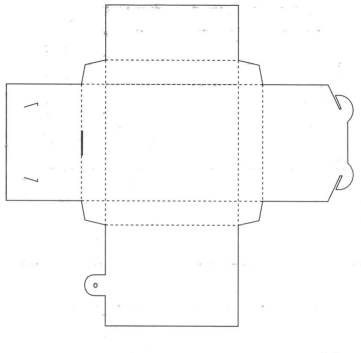

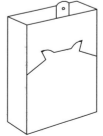

Ear Hook Lock Back Panel with Hanging Loop

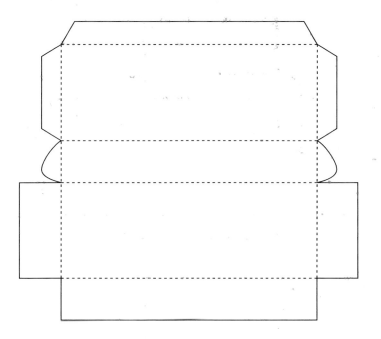

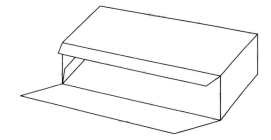

Sleeve for Cigarette Cartons 147

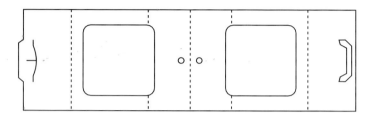

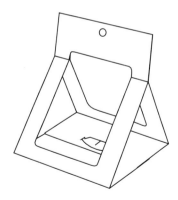

Open Sleeve for Cylindrical Products

Folders and Wrap-around Packaging

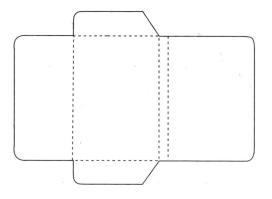

Simple Folder

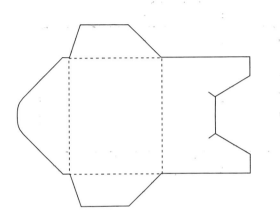

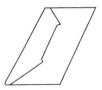

Simple Folder

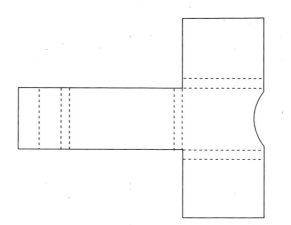

One Piece Folder

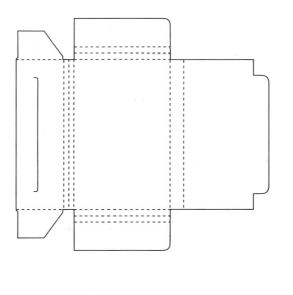

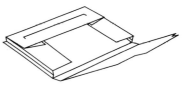

Two Section Folder 153

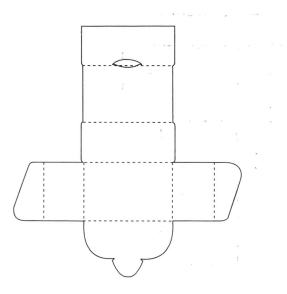

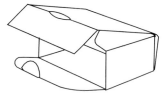

One Piece Wrap with Tab lock

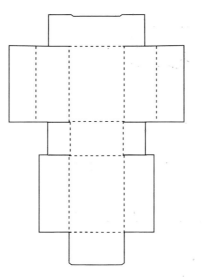

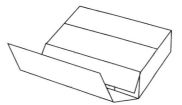

One Piece Wrap

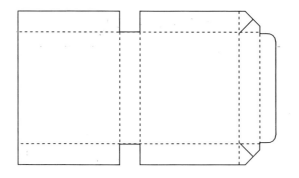

Side Seal Wrap with Tuck and Bellow

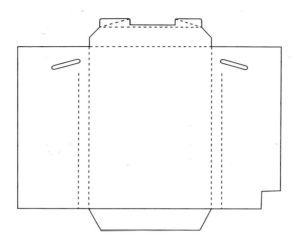

Heavy Duty Folder with Tab Lock and Expandable Sides 157

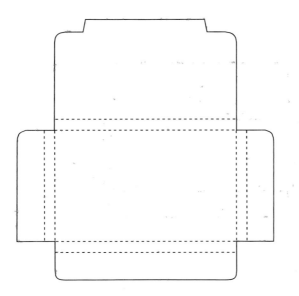

Sleeve with Glued Corners and Snap Lock Lid

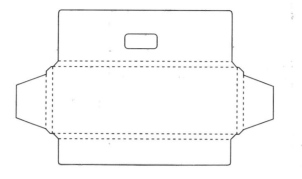

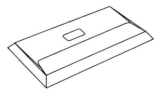

Simple Folder with Window

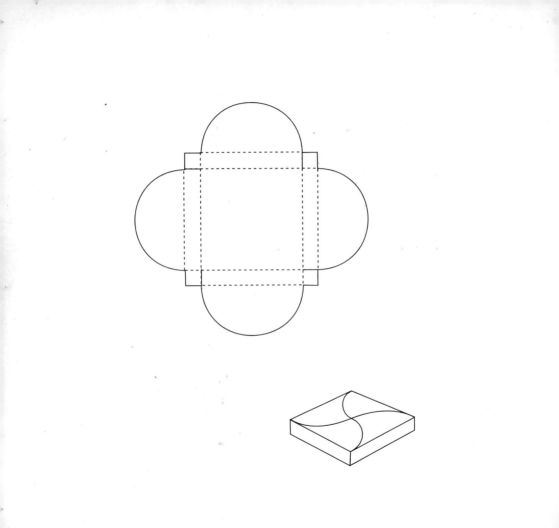

Wrap-around with Interlocking Flaps

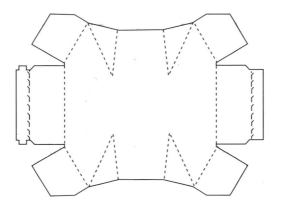

Wrap-around with Rounded Top and Zipper Bottom

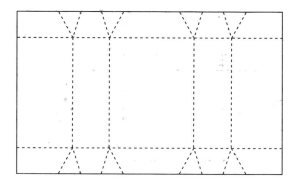

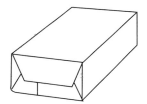

Soap Bar Wrapper

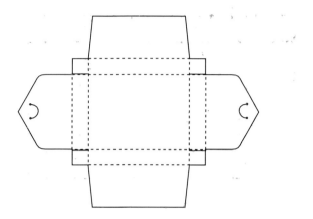

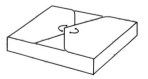

Simple Folder with Hooks for String Closure

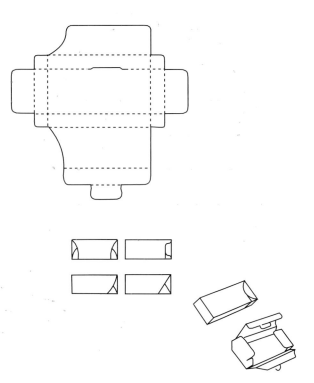

Wrap-around with Side Opening

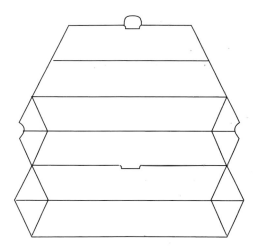

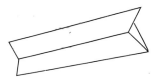

Carrier Bags and Bag-type Boxes

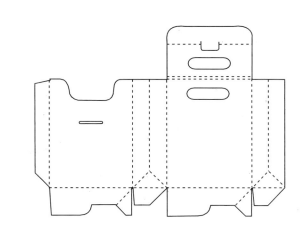

Bag with Auto Lock Bottom, Gusseted Sides and Snap Lock Closure

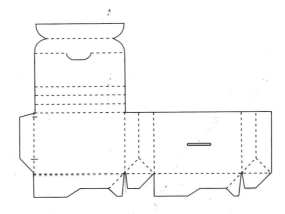

Pouch with Auto Lock Bottom, Gusseted Sides and Snap Lock Closure   169

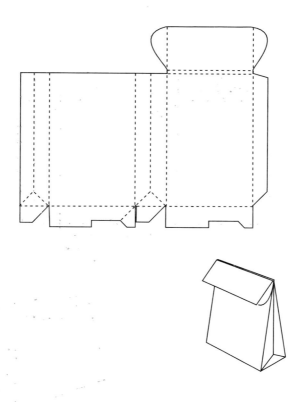

Gusseted Bag with Auto Lock Bottom and Lock-in Cover Flaps

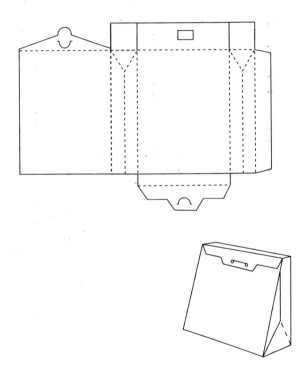

Bag with Bottom Lock, Side Gussets and Hook Lock

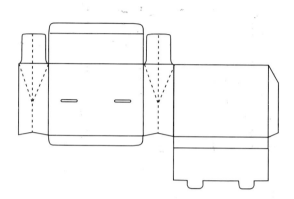

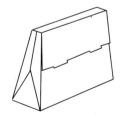

Bag with Reverse Tuck Locking Sides and Gusseted Tuck-in Closure

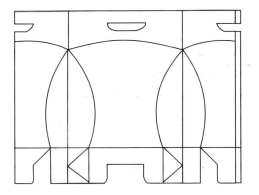

Bag with Semi Lock Bottom

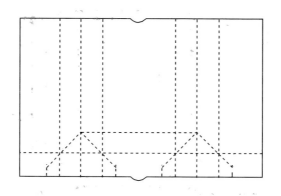

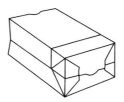

Flat Bottom Grocery Bag with Gusseted Sides

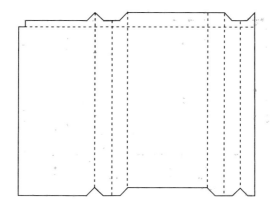

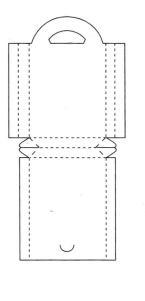

Bag with Glued Gusseted Sides, Handle and Closure Tab

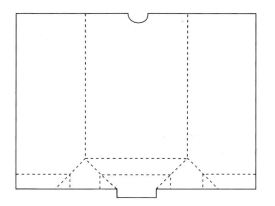

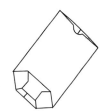

Standing Pouch with Hexagonal Bottom

Corrugated and Heavy Duty Boxes

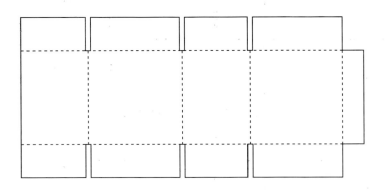

Universal Slotted Container

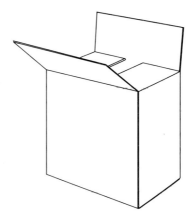

Full Overlap Slotted Container

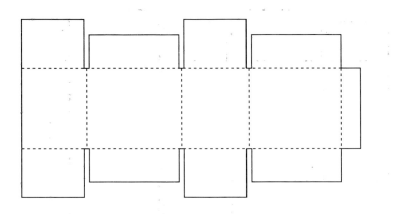

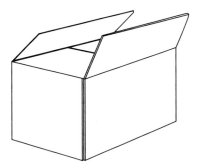

Slotted Container

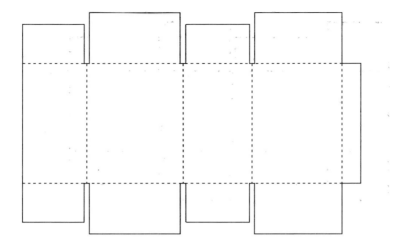

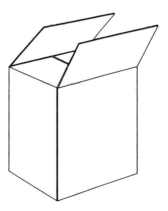

Full Overlap Slotted Container

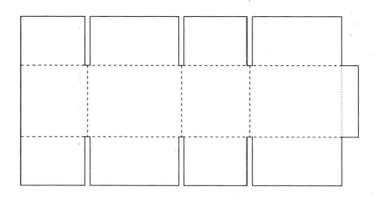

Partial Overlap Slotted Container

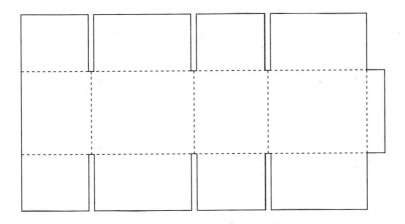

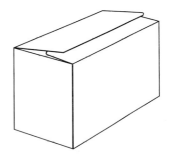

Partial Overlap Slotted Container   185

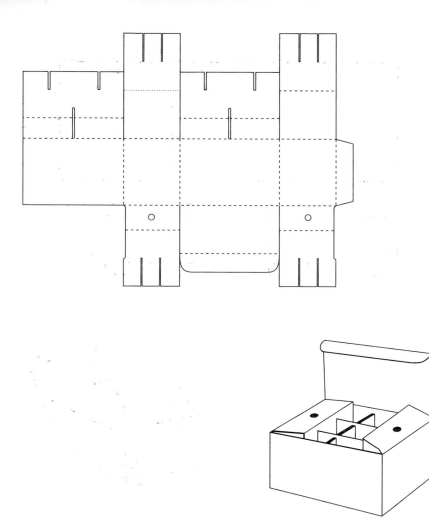

One Piece Box with Built-in 12-Cell Partition

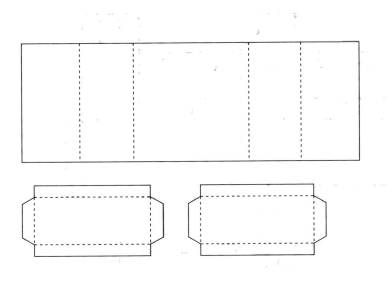

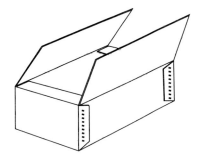

2-Bliss box

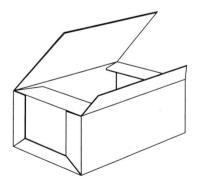

4-Bliss Box

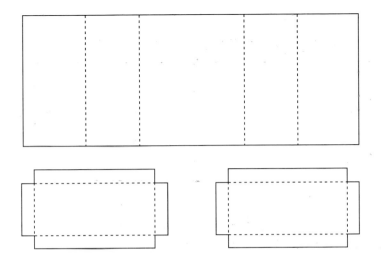

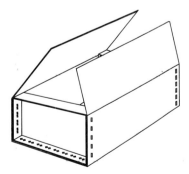

Recessed End Box

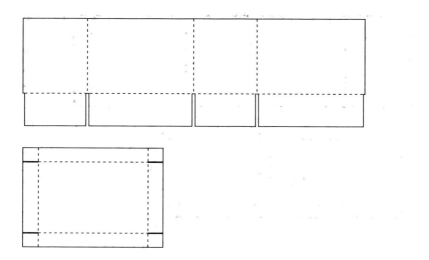

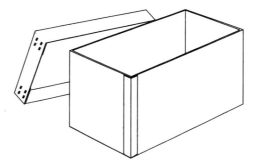

Half Slotted Box with Stapeled Cover

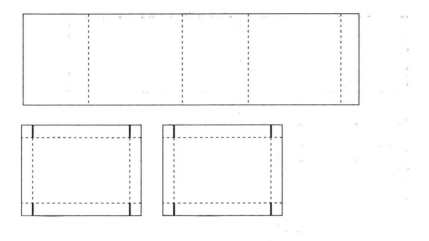

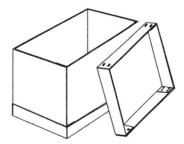

Double Cover Box

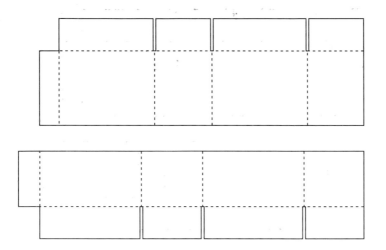

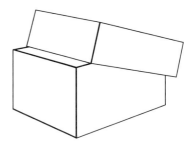

Half Slotted Box with Partial Telescopic Cover

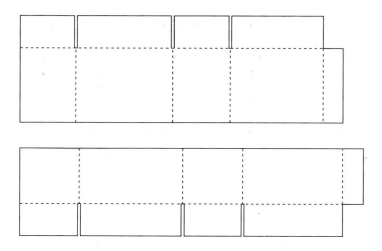

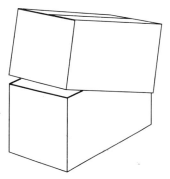

Half Slotted Box with Full Telescopic Cover

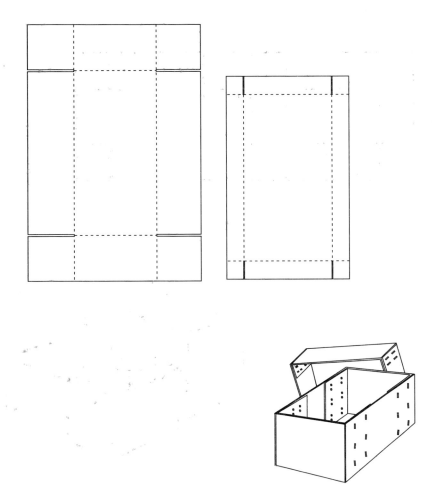

Heavy Duty Stapeled Box with Cover

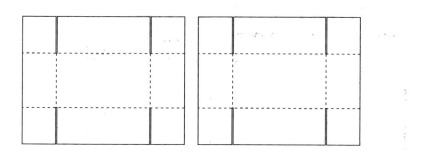

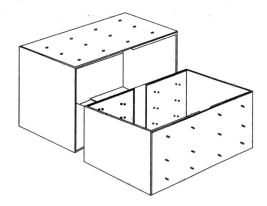

Telescopic Heavy Duty Stapeled Box

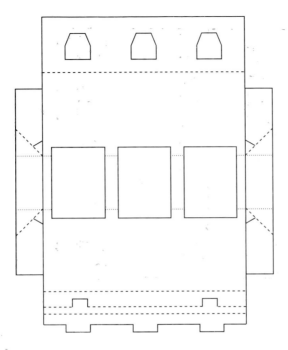

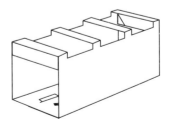

Three-Pack Ace with Lock Bottom

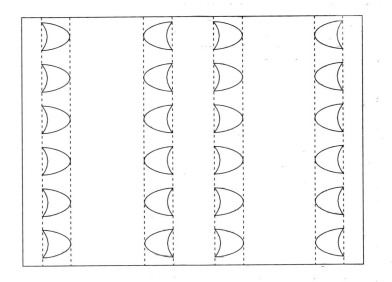

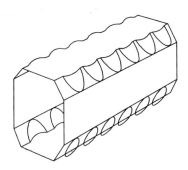

Twelve-Pack Sleeve with Cap and Heel Openings 197

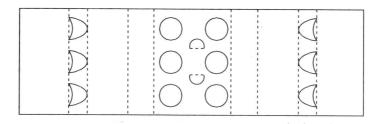

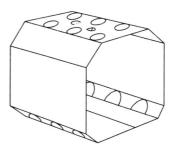

Bottom Glued Six-Pack with Neck and Heel Openings

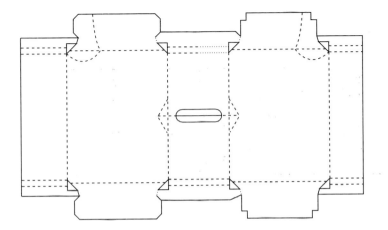

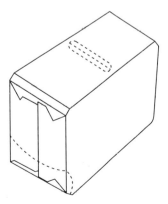

Twelve-Pack Can Carrier with Perforated Tear-Open Strip 199

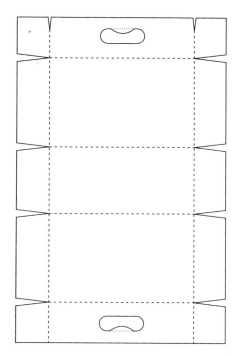

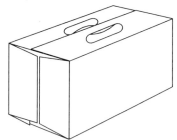

End Loading Tube with Reinforced Handle

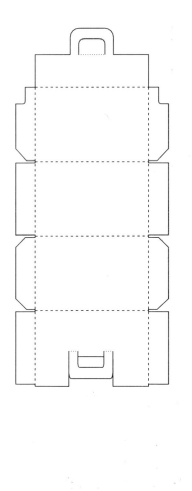

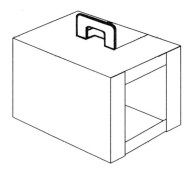

Open Ended Carrier

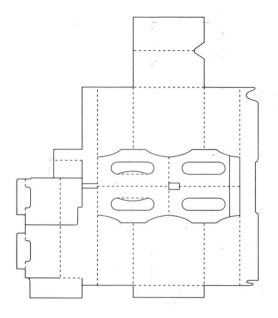

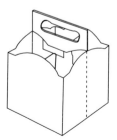

Four-Pack Bottle Carrier

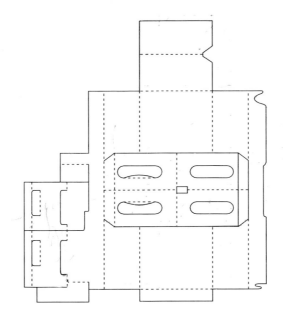

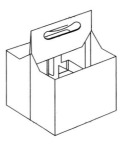

Four-Pack Bottle Carrier

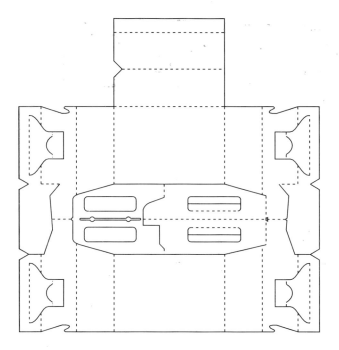

Six-Pack Bottle Carrier with Glued Bottom

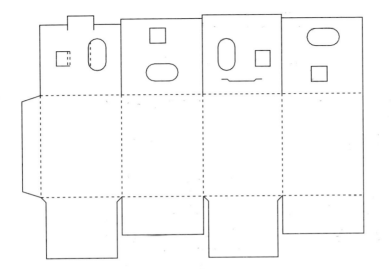

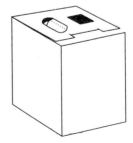

Bottom Glued Box with Carrying Holes

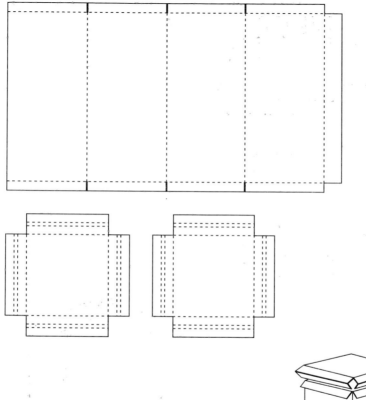

Interlocking Double Cover Box

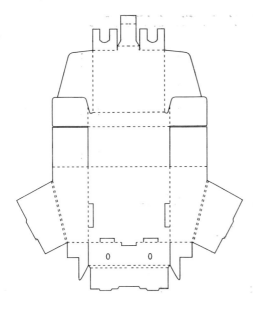

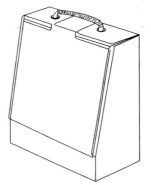

Box with Faceted Front and Handle

Heavy Duty Shippers

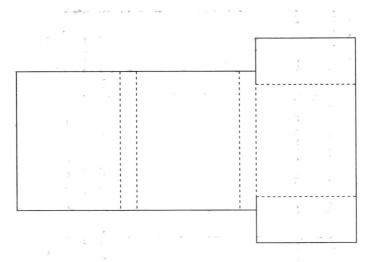

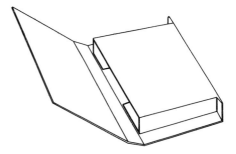

Corrugated Book Shipper with Recessed End Panel

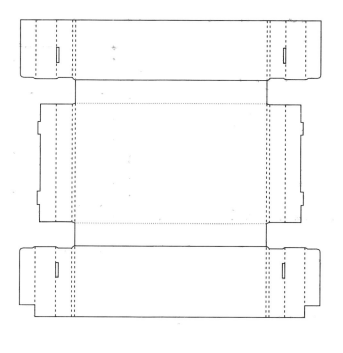

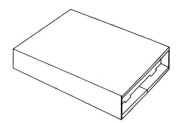

Corrugated Book Shipper with Recessed Self Locking Closures   211

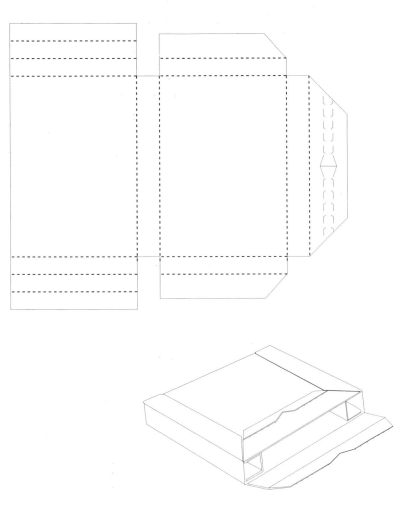

Book Shipper with Zipper and Built-in Pads

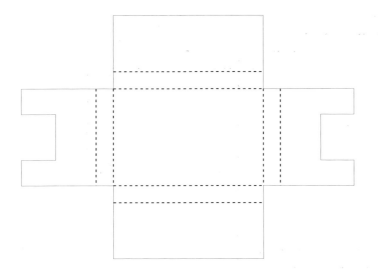

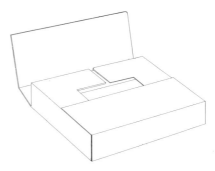

Audio Tape Shipper with Die-Cut Opening

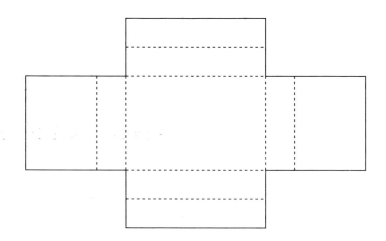

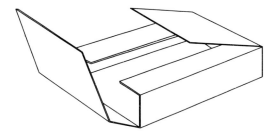

One Piece Folder

Fitments and Spacers

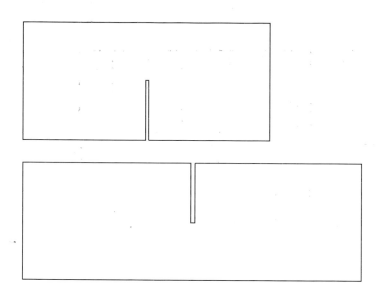

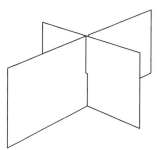

Four-Cell Slotted Partition

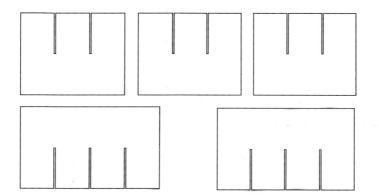

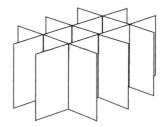

Twelve-Cell Slotted Partition

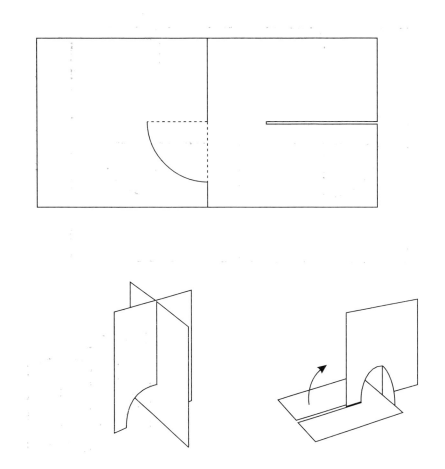

One Piece Four-Cell Partition

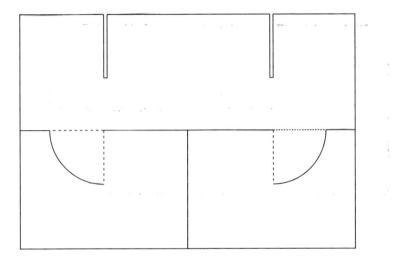

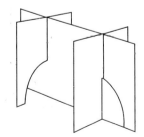

One Piece Six-Cell Partition

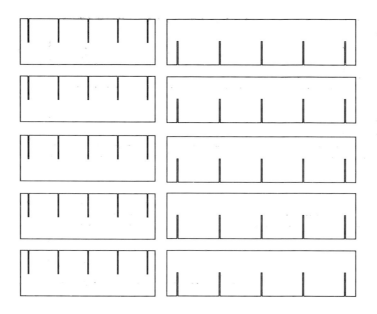

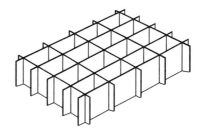

16-Cell Extension Partition

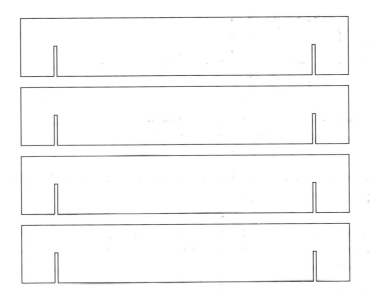

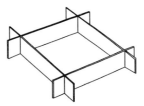

Single Cell Extension Partition

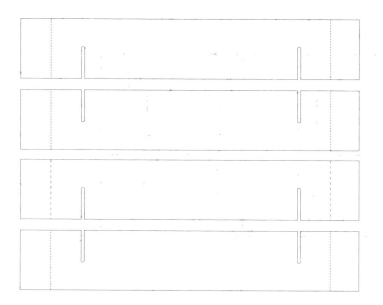

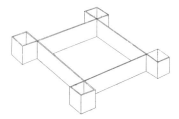

Single Cell Support Partition

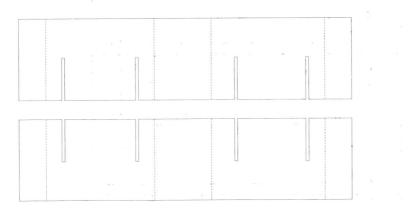

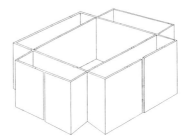

Scored and Slotted Compartment Filler

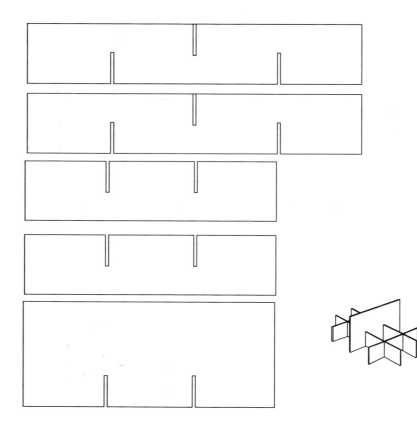

Twelve-Cell with High Divider

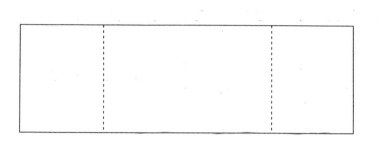

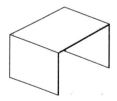

'C' Partition

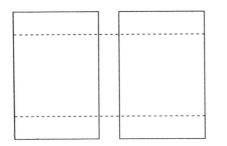

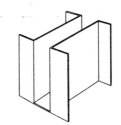

'U' Partition

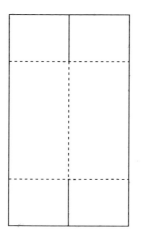

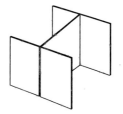

'I' Beam Partition

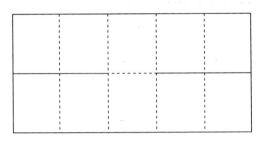

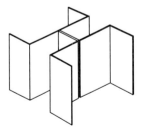

'H' Beam Partition with Wings

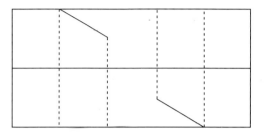

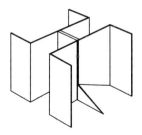

'H' Beam Partition with Feet

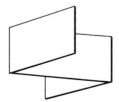

'Z' Partition

Double 'Z' Partition

Triple 'Z' Partition

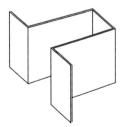

Modified 'T' Partition

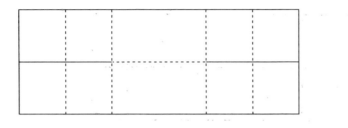

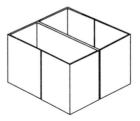

Linear 'H' Partition

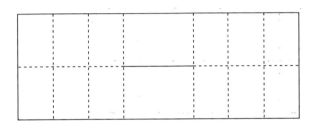

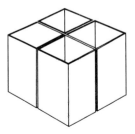

Folded Four-Cell Partition

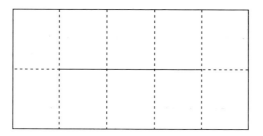

Folded Five-Cell Partition

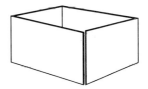

Open Liner

Open Liner

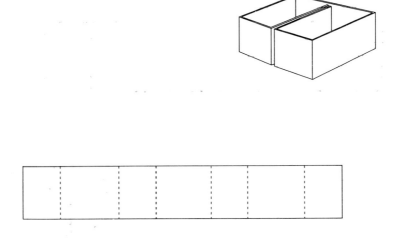

Open Compartment Liner

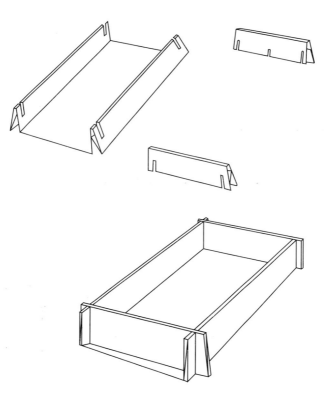

Open Compartment Liner

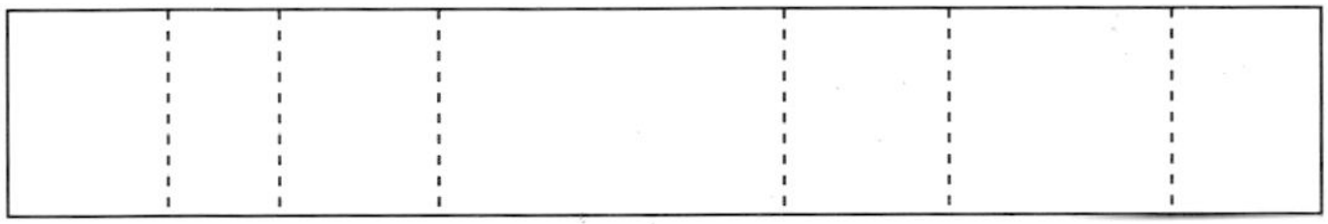

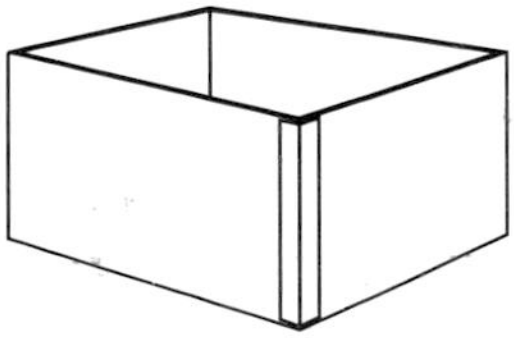

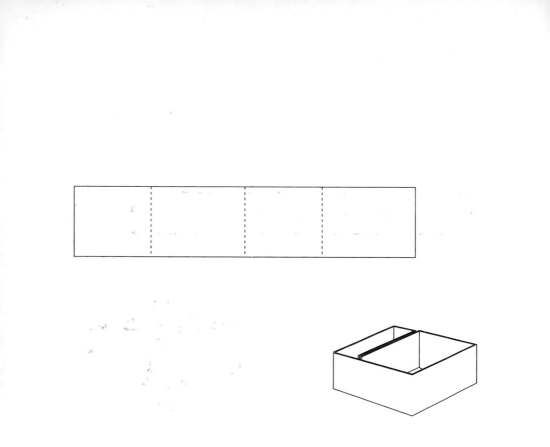

Open Compartment Liner

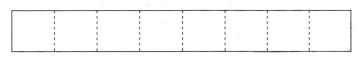

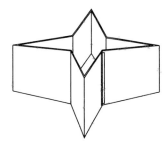

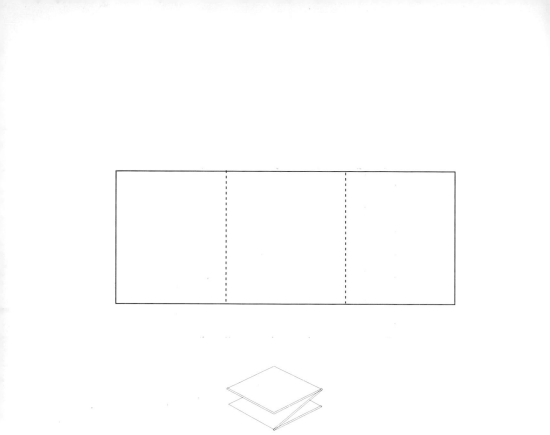

Spacer

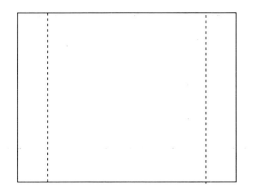

Cushion Pad

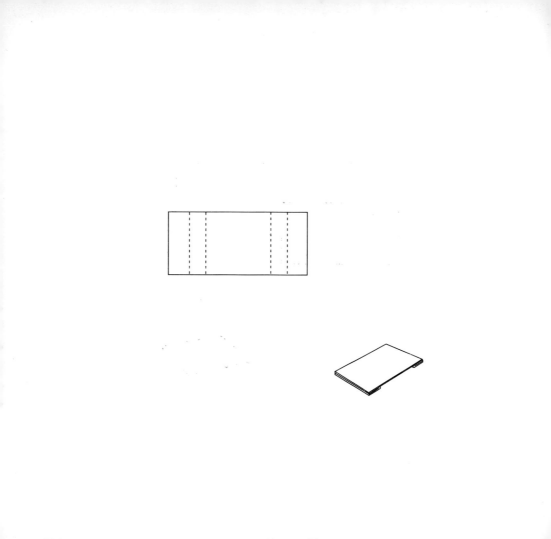

Cushion Pad

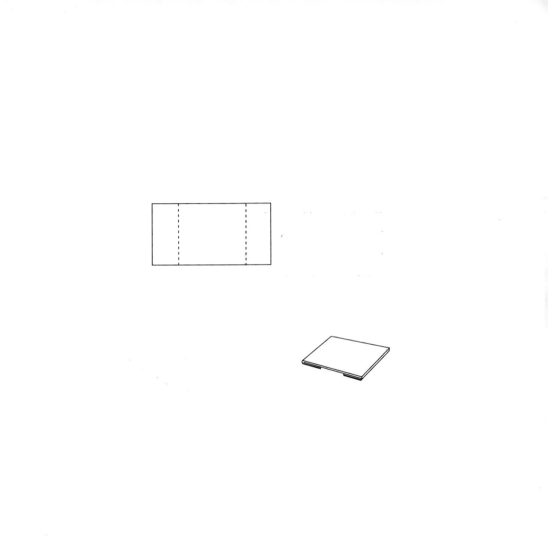

Cushion Pad

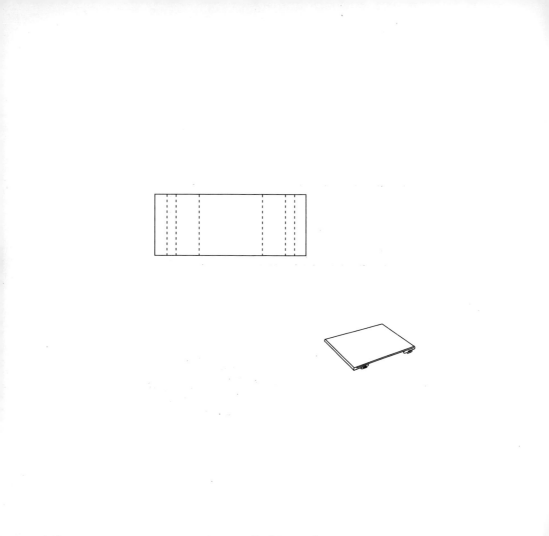

Cushion Pad

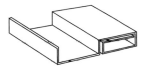

Cushion Pad

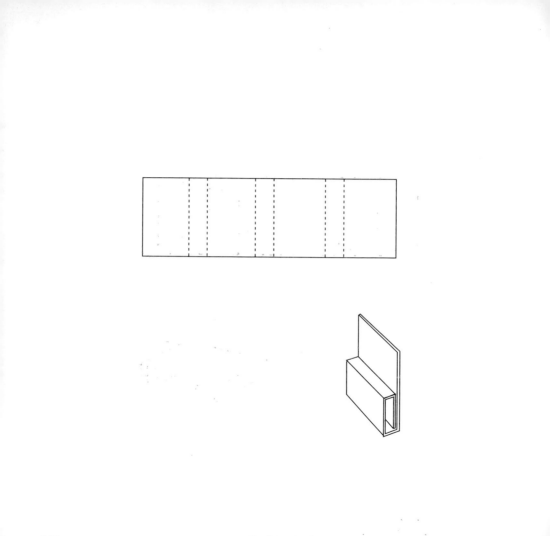

Cushion Pad

Cushion Pad

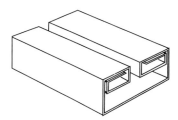

Cushion Pad

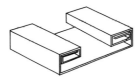

Cushion Pad

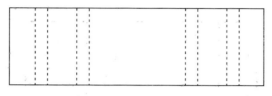

Cushion Pad

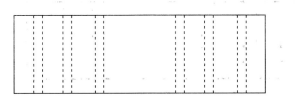

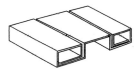

Cushion Pad

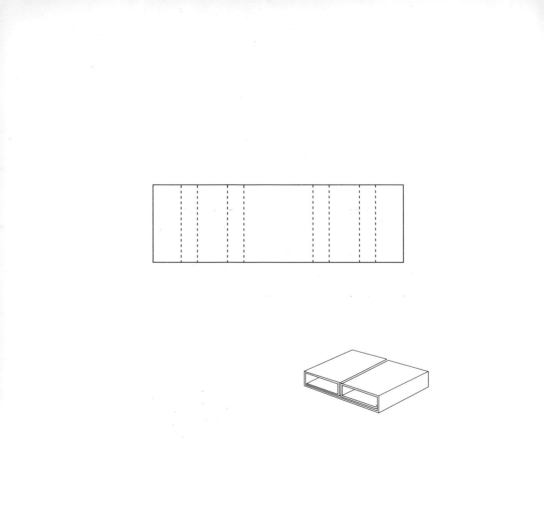

Cushion Pad

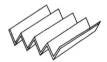

Accordion Brace Pad

Clearance Pad

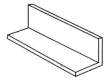

Corner Protection

Scored Sheets

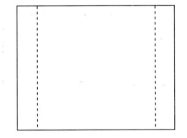

Scored Sheet

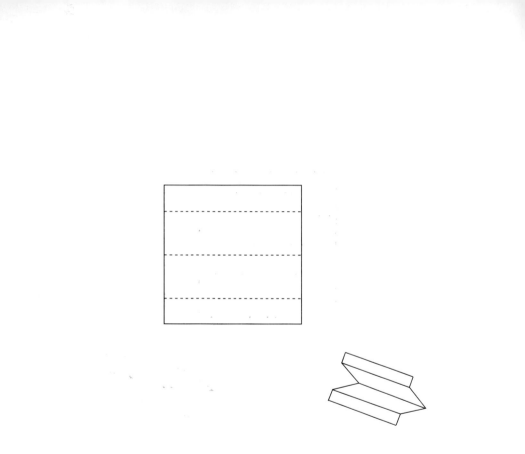

Brace Pad

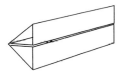

Corner Brace Pad

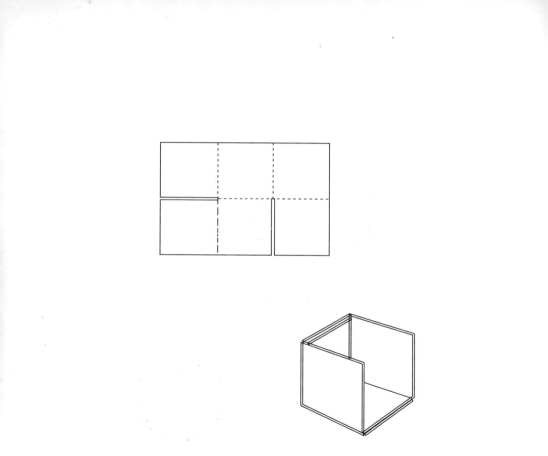

Slotted Corner Protector

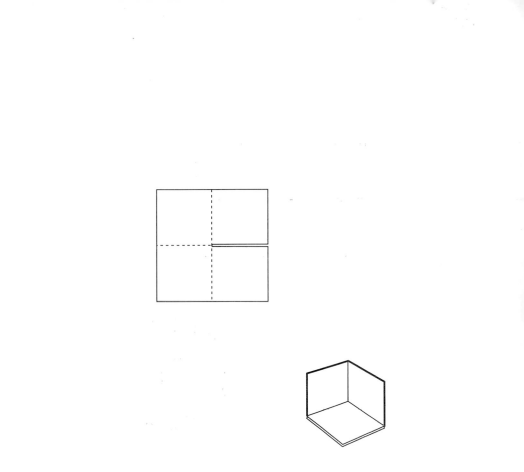

Slotted Corner Protector

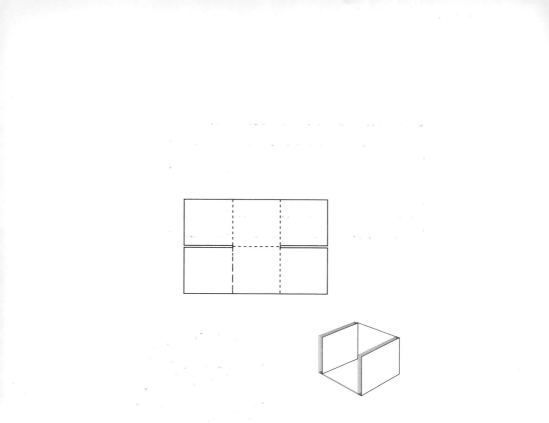

Slotted Corner Protector

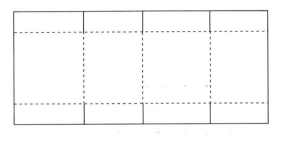

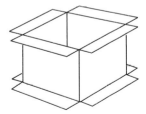

Liner with Folded Sides

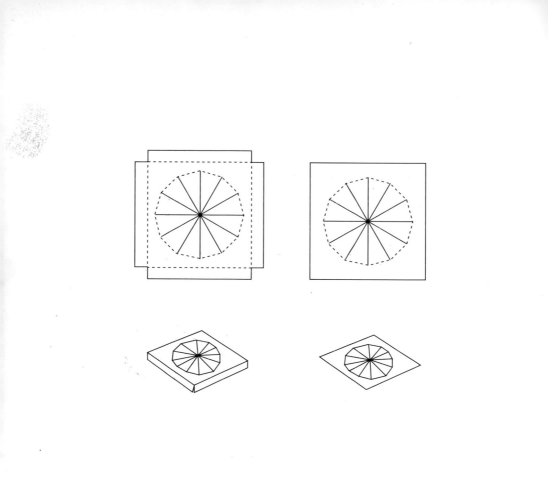

Die-Cut Pad and Tray

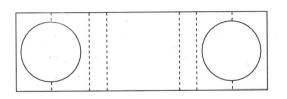

Die-Cut Anchor Pad

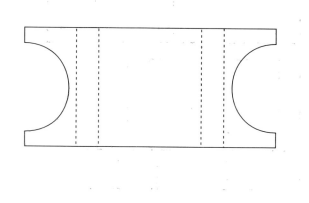

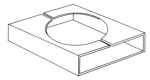

Die-Cut Anchor Pad

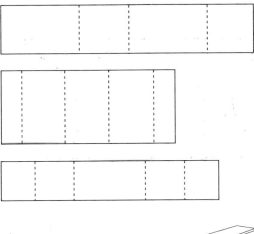

Double Lined Slide Box

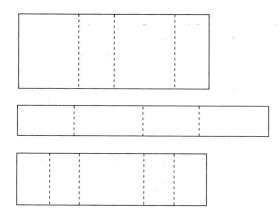

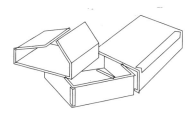

Triple Slide Box

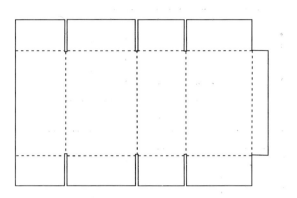

Wrap-around Blank

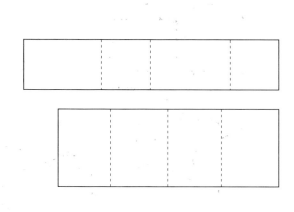

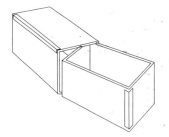

Double Slide Box

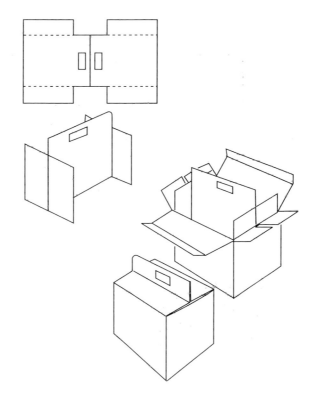

Partition with Handle

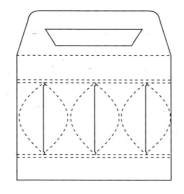

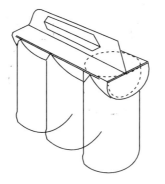

Spacer/Handle for Shrink Wrap Packing

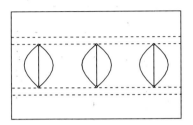

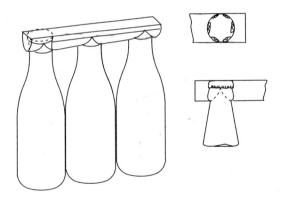

Spacer/Handle for Shrink Wrap Packing

Creative Forms

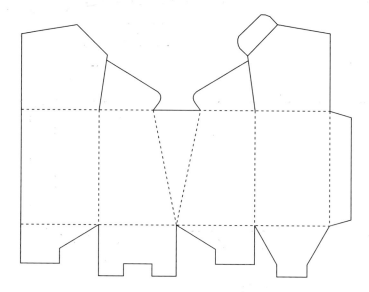

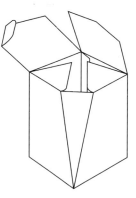

Tapered Facet on One Side

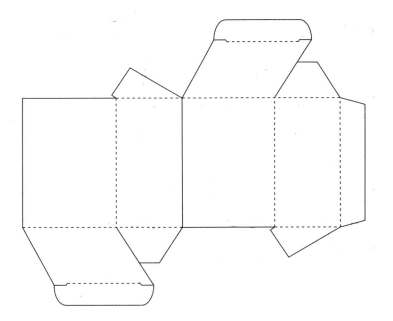

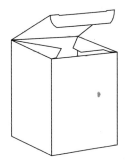

Trapezoid Package with Reverse Tucks

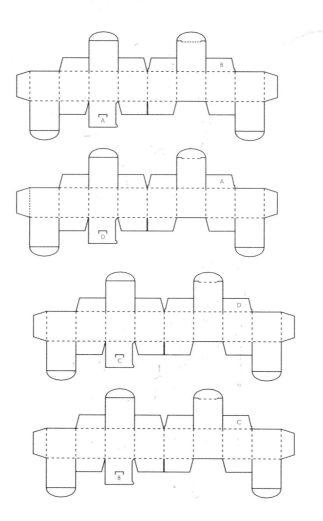

Eight Cubes Interlocking at the Indicated Dust Flaps

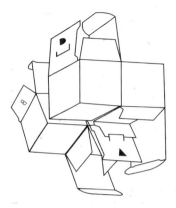

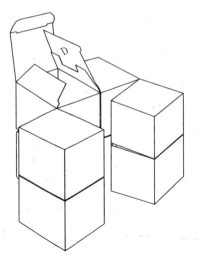

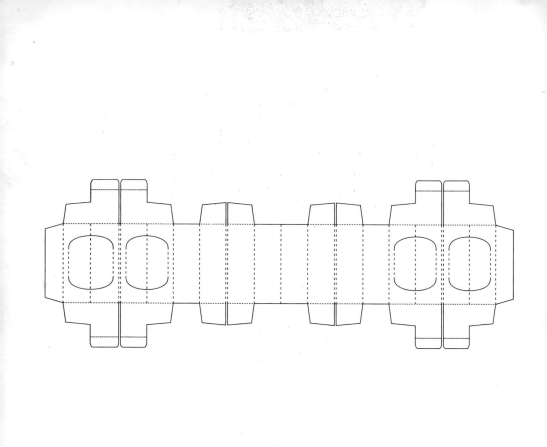

Quadruple Window Display Box

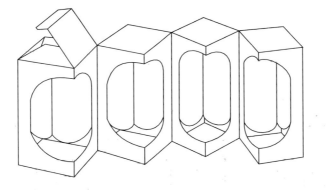

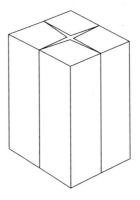

Quadruple Window Display Box

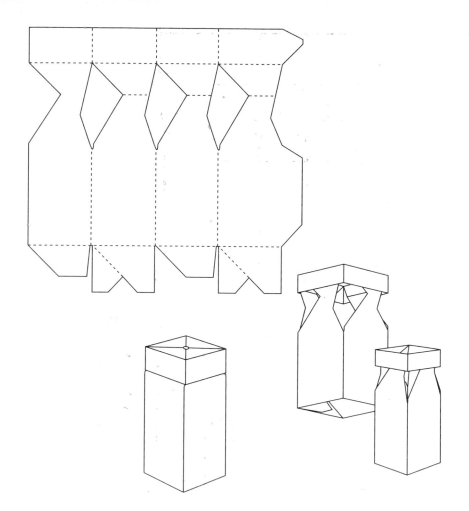

Slide Down Ring to Open

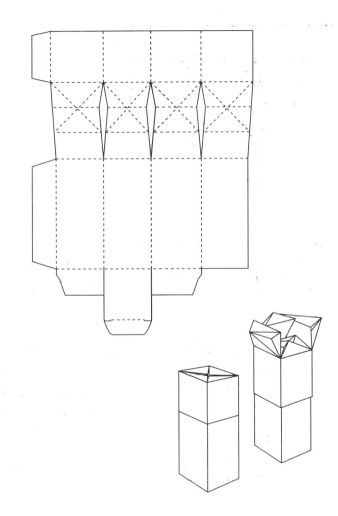

Slide Down Ring to Open

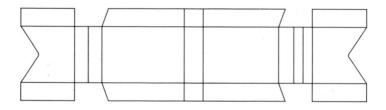

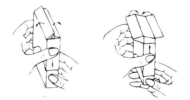

Slide Down to Open

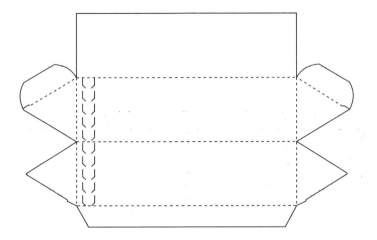

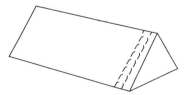

Triangle with Tuck Closures and Tear-away Zipper 289

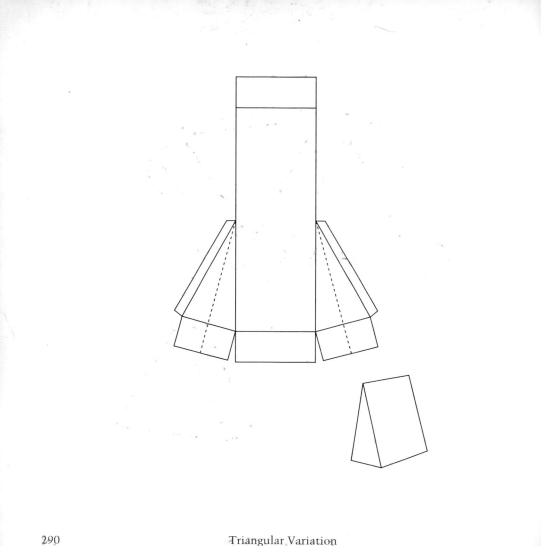

Triangular Variation

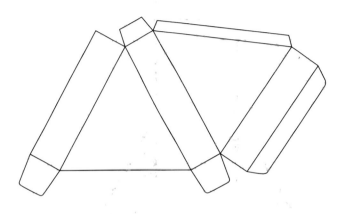

Triangular Variation

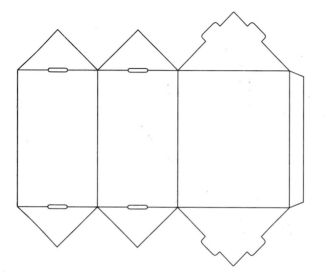

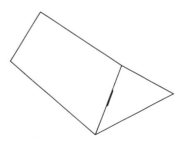

Triangular Variation

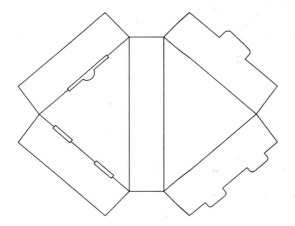

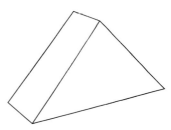

Triangular Variation

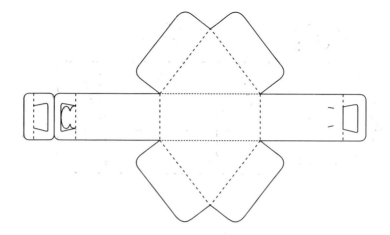

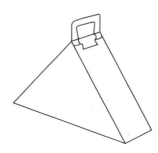

Triangular Box with Handle

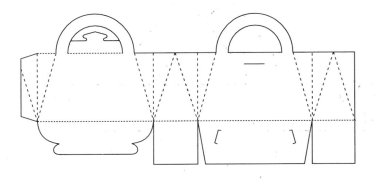

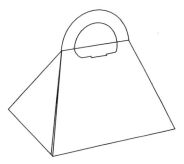

Bag with Hook Lock Bottom and Tab Lock Top 295

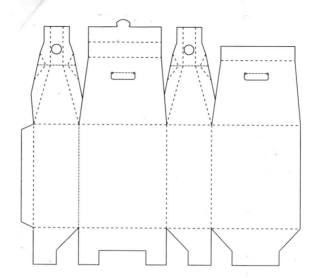

Semi Lock Bottom and Tapered Top with Handle

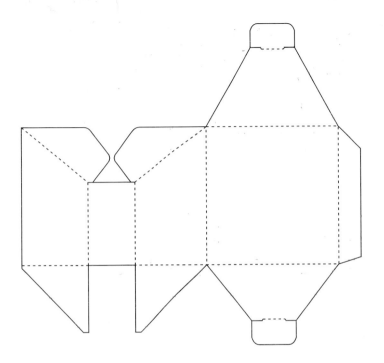

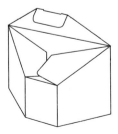

Box with Tapered Front

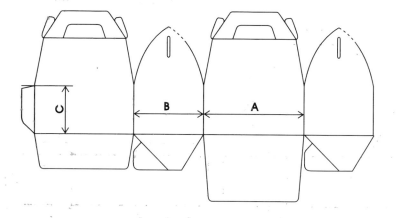

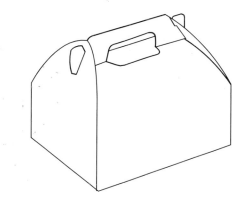

Bag with Tapered Front and Handle

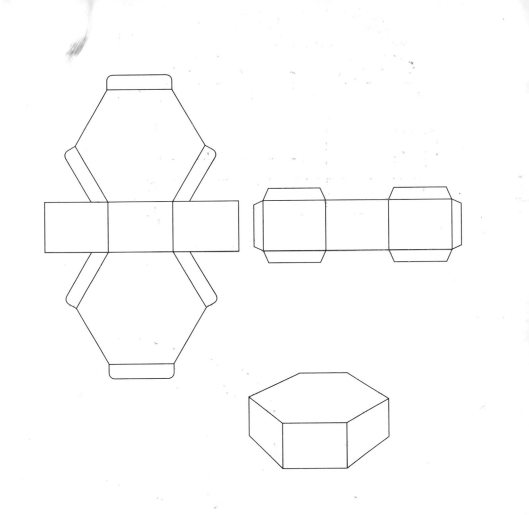

Hexagonal Ring Box

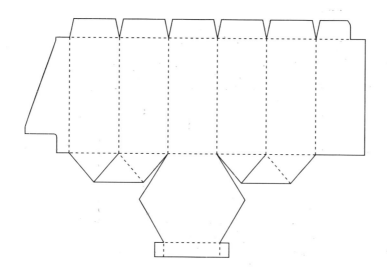

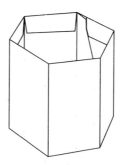

Hexagonal Sleeve with Auto Lock Bottom

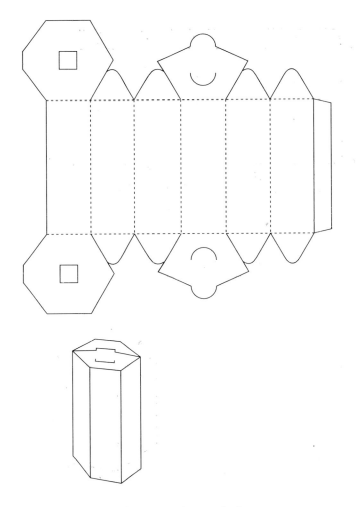

Hexagon with Snap lock

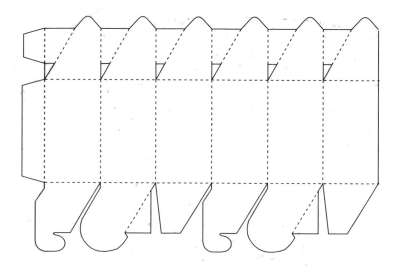

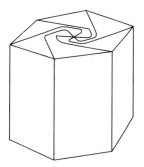

Hexagonal Carton with Push-in Closure and Auto Lock Bottom

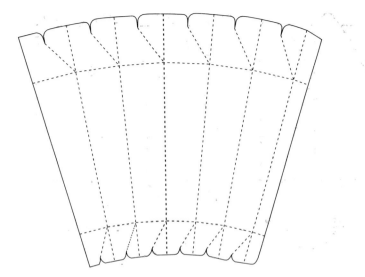

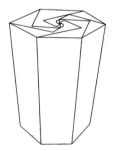

Hexagonal Tapered Box with Push-in Top and Bottom

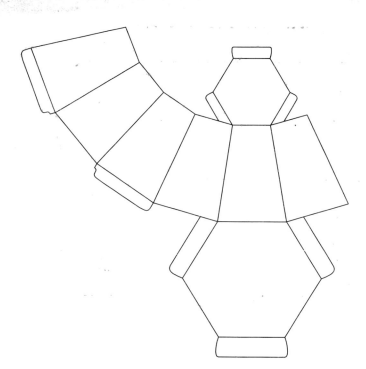

Tapered Hexagonal Container

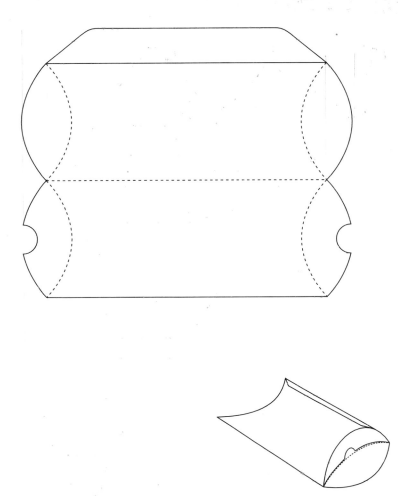

Pillow Pack

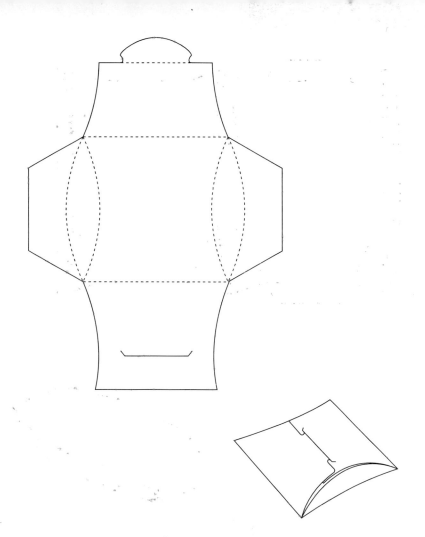

Pillow Pack with Tuck

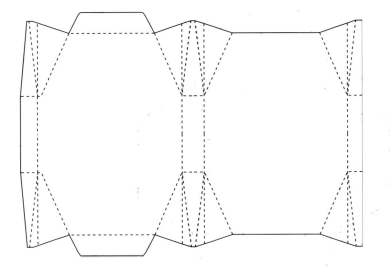

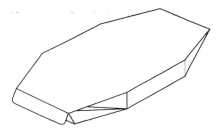

Wide Mouth Box

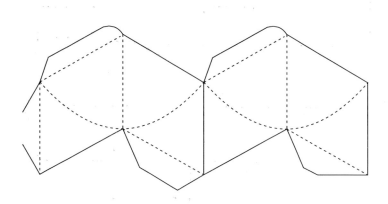

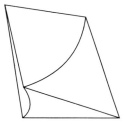

Straight and Curved Creasing

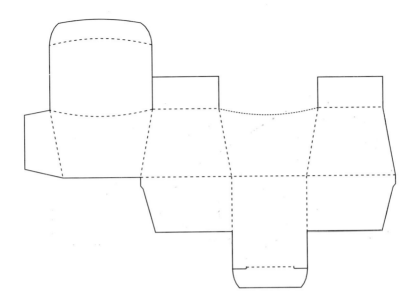

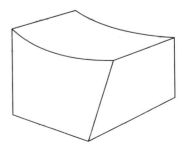

Reverse Tuck Curved Carton

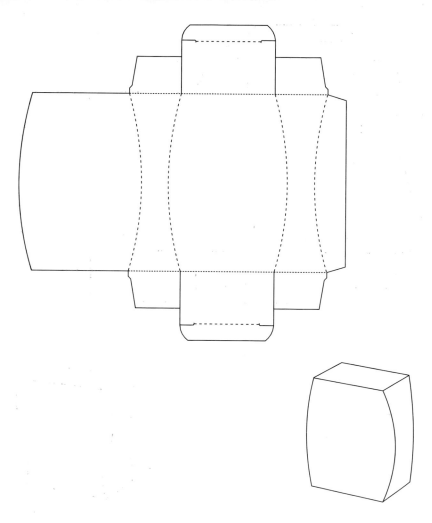

Straight Tuck Carton with Curved Sides

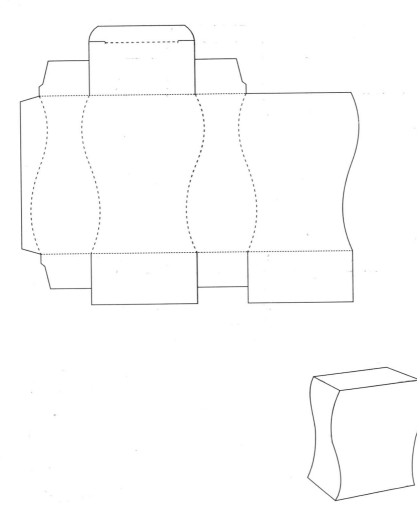

Double Curved Carton

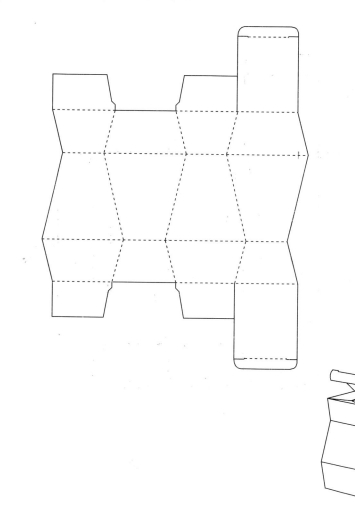

Triple Tapered Box

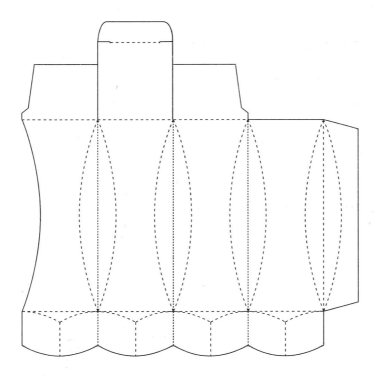

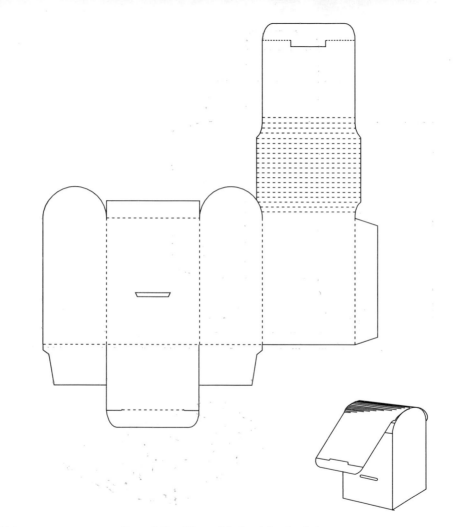

Round Top Flap with Straight Tuck Bottom

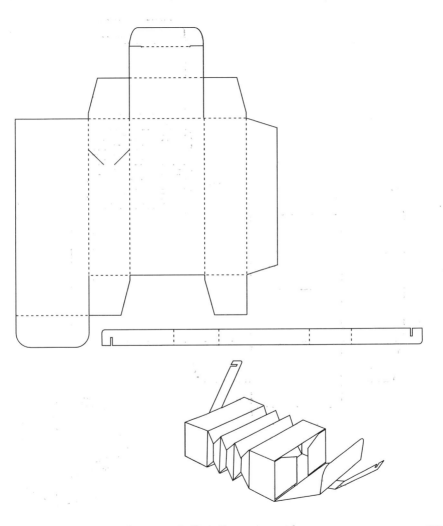

Bellow Box

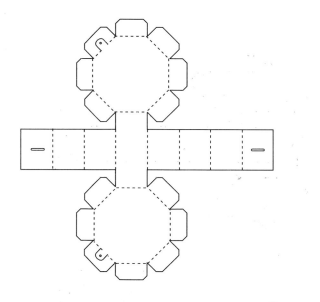

Pentagonal Box

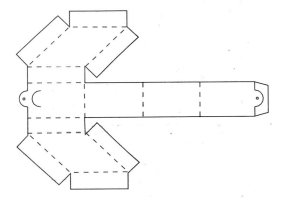

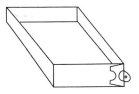

Diamond Shape Tray

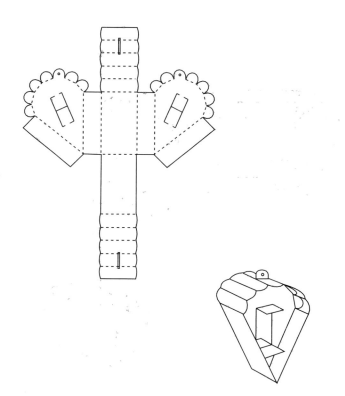

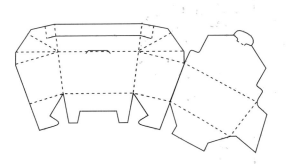

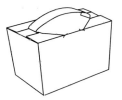

Box with Tapered Sides, Snap Lock Bottom and Handle   319

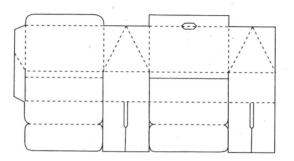

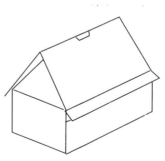

House Shape Package

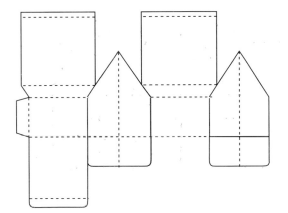

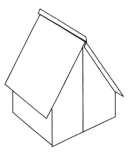

House Shape Package

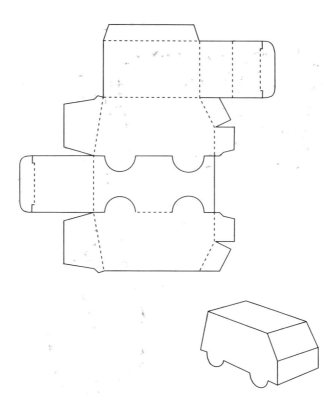

Car Shape Box

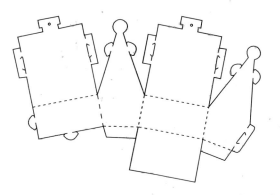

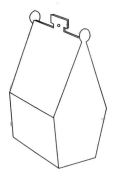

Church Shape Box

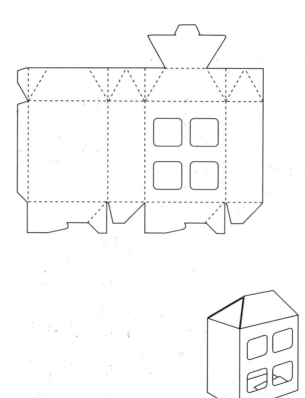

Two Storey Bungalow

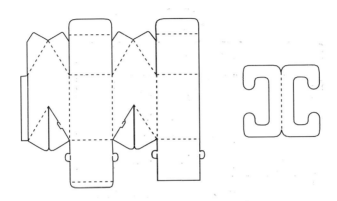

Tower Maker

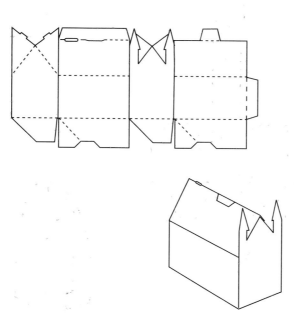

House Shape Box

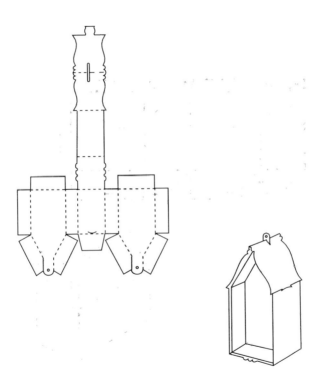

Cuckoo Clock

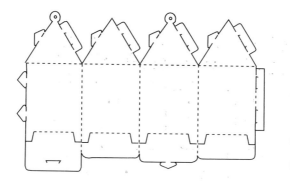

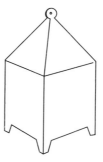

Hut Shape Box

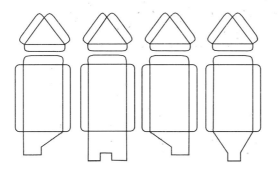

Pagoda Box

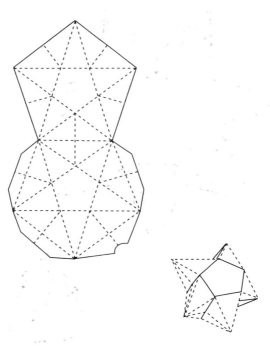

Star Shape Box

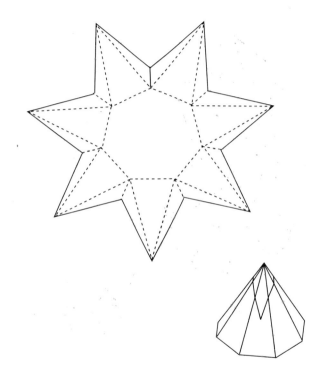

Conical Box

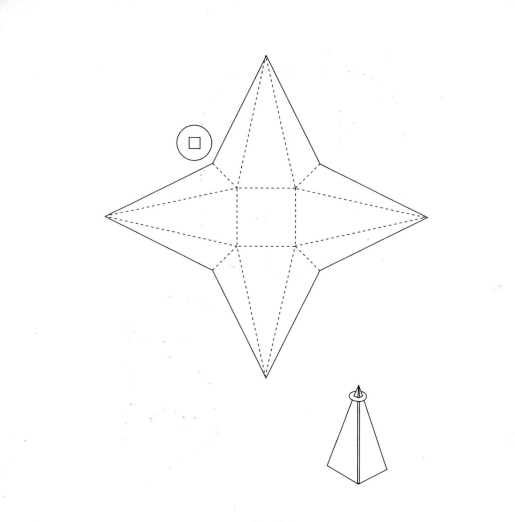

Pyramid Box

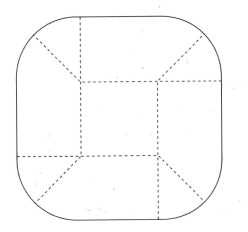

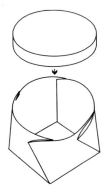

Basket with Lid

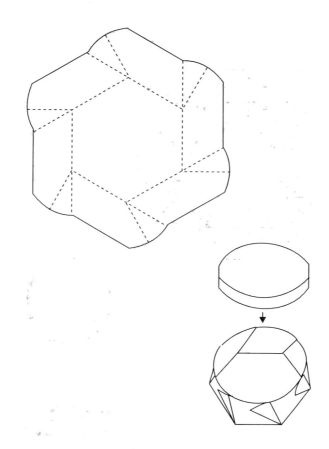

Round Basket with Lid

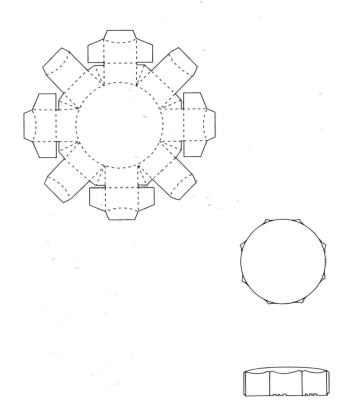

Octagonal Tray

Counter Displays

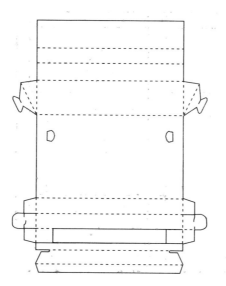

Cosmetics Display

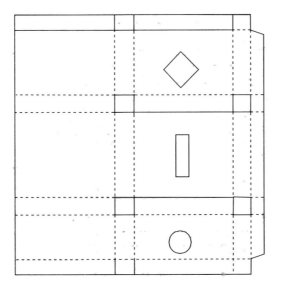

One Piece Open View Display Box

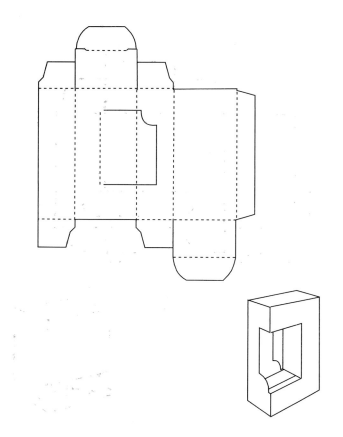

Reverse Tuck with Die-Cut Window

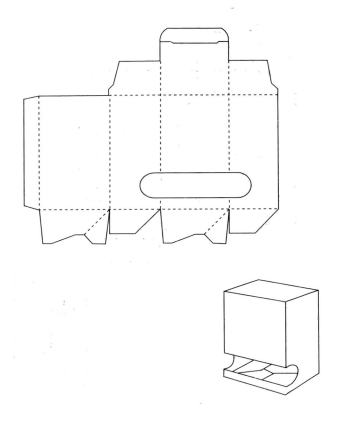

Auto Lock Bottom with Gravity-fed Dispenser

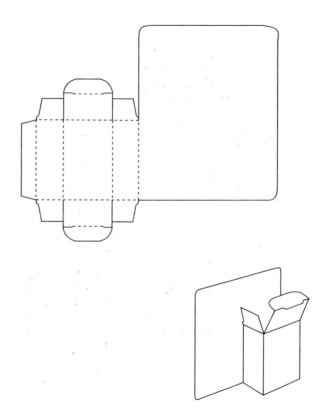

Straight Tuck with Back Drop Panel

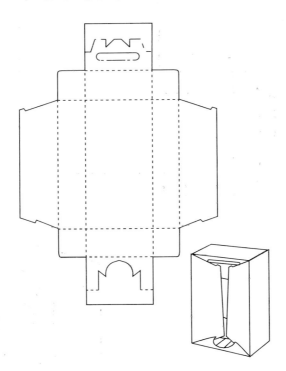

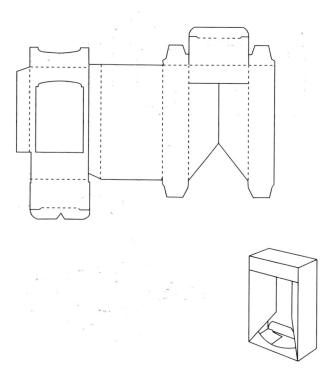

Display Box with Window

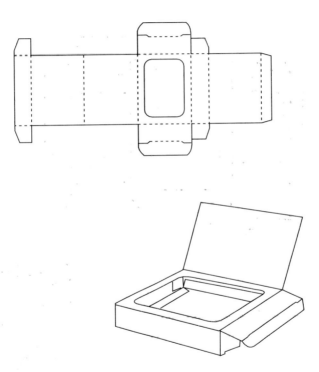

Double Cover Box

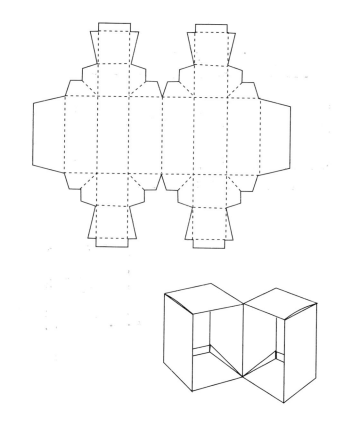

Two Compartment Display Box

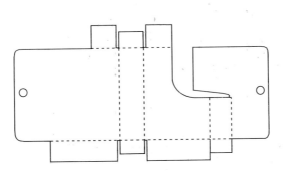

Display Pack with Hanging Panel

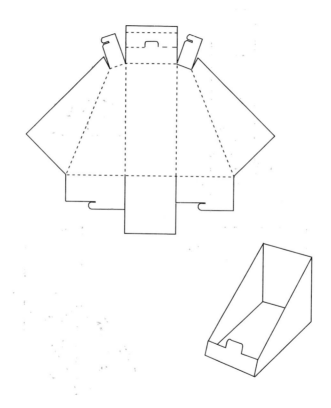

Counter Display with Double Walls

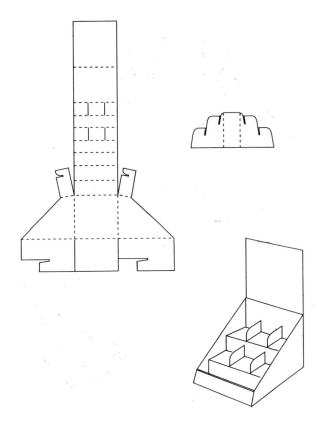

Counter Display with Steps and Dividers

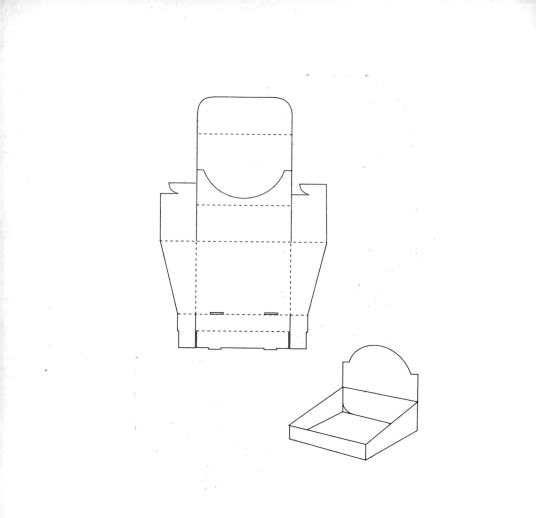

Hook lock and Walker Lock Combination Display

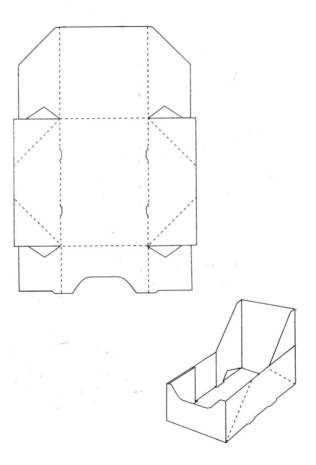

Folding Counter Display with Bottom Lock Slots 351

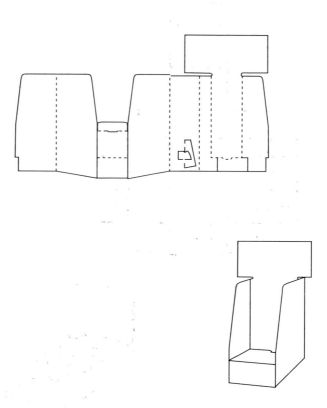

Counter Display with Built-in Easel

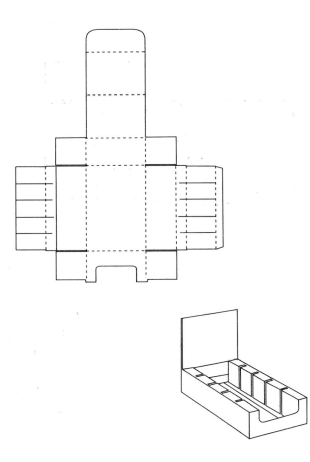

Display Pack with Double Side Walls

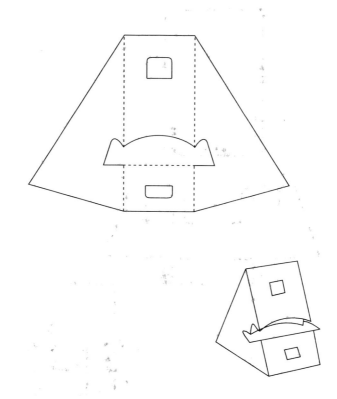

Double Wing Easel Display

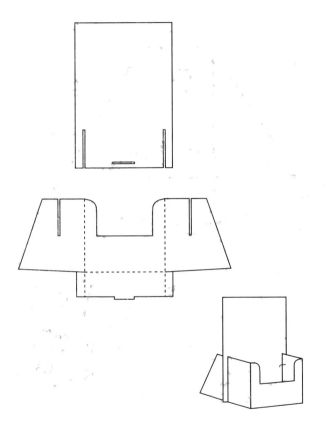

Two Piece Brochure Display

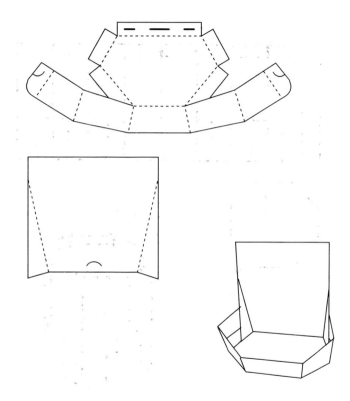

Two Piece Basket and Back Panel Counter Display

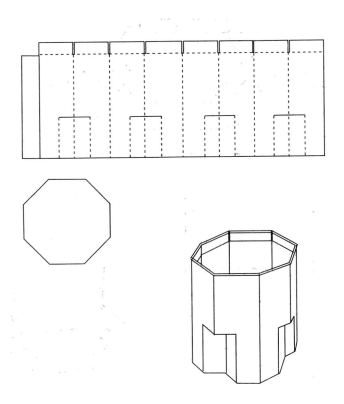

Corrugated Floor Display

Set-up Boxes

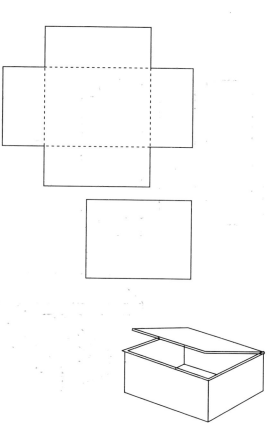

Plain Hinged Lid Box

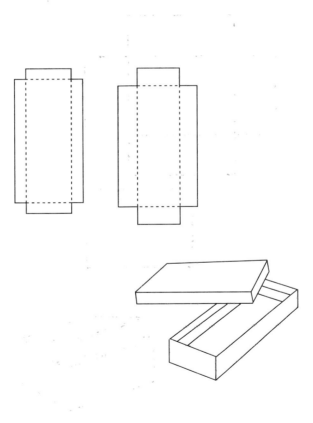

Tray with Partial Telescopic Lid

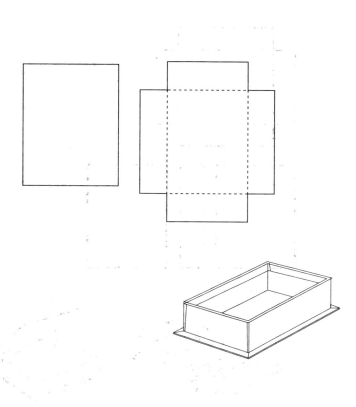

Box with Extending Bottom Panel

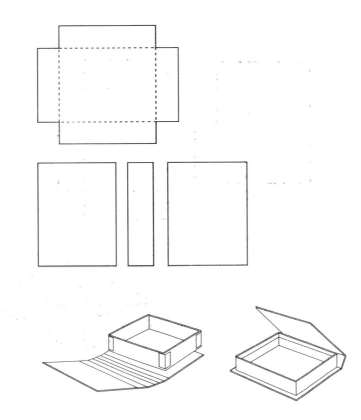

Box with Extending Bottom Panel and Lid

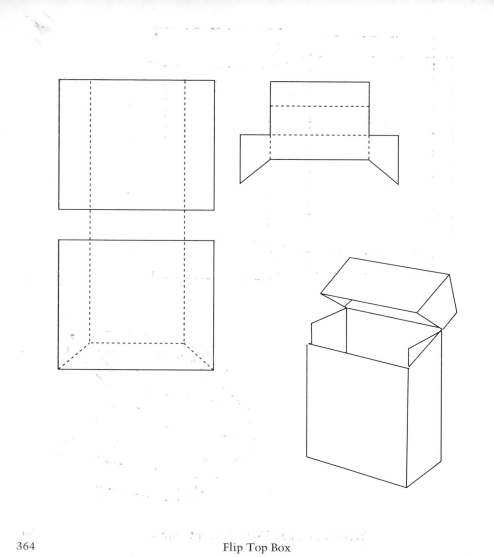

Flip Top Box

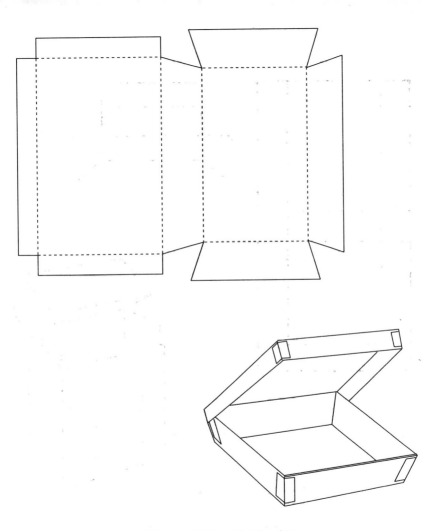

Tapered Tray and Lid with Glued Corners

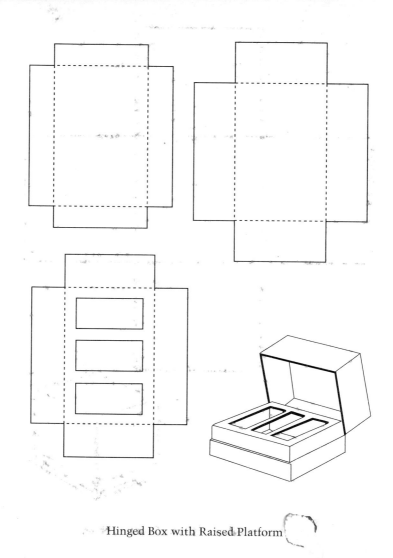

Hinged Box with Raised Platform

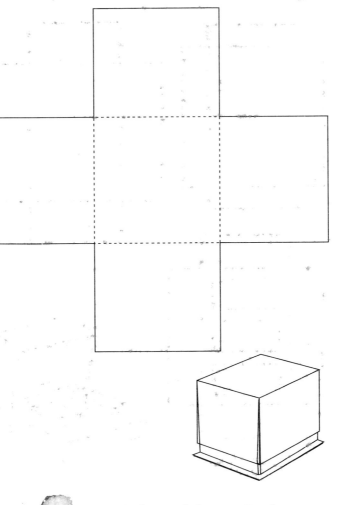

copic Box with Extended Bottom Panel

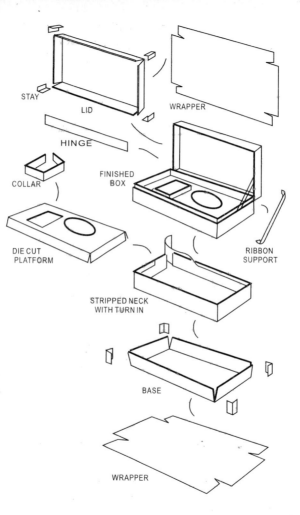

STAY

LID

WRAPPER

HINGE

COLLAR

FINISHED BOX

DIE CUT PLATFORM

RIBBON SUPPORT

STRIPPED NECK WITH TURN IN

BASE

WRAPPER

Set-up Box Construction